T0250720

# Lecture Notes in Computer Science 738

Edited by G. Goos and J. Hartmanis

Advisory Board: W. Brauer   D. Gries   J. Stoer

Matthias Weber   Martin Simons
Christine Lafontaine

# The
# Generic Development
# Language Deva

## Presentation and Case Studies

Springer-Verlag
Berlin Heidelberg New York
London Paris Tokyo
Hong Kong Barcelona
Budapest

Series Editors

Gerhard Goos
Universität Karlsruhe
Postfach 69 80
Vincenz-Priessnitz-Straße 1
D-76131 Karlsruhe, Germany

Juris Hartmanis
Cornell University
Department of Computer Science
4130 Upson Hall
Ithaca, NY 14853, USA

Authors

Martin Simons
Matthias Weber
Institut für angewandte Informatik, Technische Universität Berlin
Franklinstraße 28-29, D-10587 Berlin, Germany

Christine Lafontaine
Unité d'Informatique, Université Catholique de Louvain Place
Sainte-Barbe 2, 1348 Louvain-La-Neuve, Belgium

CR Subject Classification (1991): D.3, F.3-4, D.1.1


ISBN 3-540-57335-6 Springer-Verlag Berlin Heidelberg New York
ISBN 0-387-57335-6 Springer-Verlag New York Berlin Heidelberg

This work is subject to copyright. All rights are reserved, whether the whole or part
of the material is concerned, specifically the rights of translation, reprinting, re-use
of illustrations, recitation, broadcasting, reproduction on microfilms or in any other
way, and storage in data banks. Duplication of this publication or parts thereof is
permitted only under the provisions of the German Copyright Law of September 9,
1965, in its current version, and permission for use must always be obtained from
Springer-Verlag. Violations are liable for prosecution under the German Copyright
Law.

© Springer-Verlag Berlin Heidelberg 1993
Printed in Germany

Typesetting: Camera-ready by author
Printing and binding: Druckhaus Beltz, Hemsbach/Bergstr.
45/3140-543210 - Printed on acid-free paper

# Foreword

The Deva endeavor is almost ten years old: the requirements for the Esprit project ToolUse (1985-1990) were discussed in late 1984. The present preface offers a nice opportunity to consider the enterprise in the light of experience. The overall problem tackled in the design of Deva can be outlined as follows; the word *proofs* denotes deductions going from hypotheses to theses or from specifications to programs. On the one hand, we are used to writing and reading *human proofs*; these are sometimes unstructured, imprecise, and even incorrect. On the other hand, some of us strive to write *formal proofs*; such proofs are often too detailed and hard to understand. It is tempting to bridge the gap between human and formal proofs by introducing *formal human proofs*, or human formal proofs if one prefers, so as to remove the shortcomings of the two modes of expression without losing their respective qualities.

This task is not as hopeless as it seems. Indeed, there is a permanent tendency to improve the style of human proofs. Sloppiness, for instance, is combatted by systematic use of consecutive formal expressions separated by careful descriptions of the laws used at each step. Composition is enhanced by nesting proofs: sub-proofs correspond to sub-blocks or lemmas, and hypotheses to declarations. Such improved human proofs could be termed *enlightening proofs*. In these, the initial laws, the consecutive propositions, and the overall structure are all formalized; only the proof steps and the scope rules remain informal. The discipline fostered by such enlightening proofs reduces the temptation to cheat in reasoning; this has been a sobering personal experience. Other efforts towards formal human proofs aim at making formal proofs more human. The corresponding techniques are effective, if not original: they include systematic composition, readable notations, and automatic sub-deductions such as pattern-matching. In spite of these varied efforts from both sides, the gap between human proofs and formal ones remains a wide one. The Deva enterprise was intended to reduce it further by humanizing formal proofs a bit more.

Deva is essentially a typed functional language. The primitive functions express proof steps. Each such step is typed by an input and an output type; this pair of types expresses the propositions connected by the step, and thus amounts to a deduction rule. This view of functions as proofs and of types as propositions has been known in logic since the sixties. It differs, however, from related approaches. Indeed, a fruitful principle in computing is to consider types as abstract values and type elaboration as an abstract computation. A straightforward consequence is to view type expressions as abstract function expressions: the syntax remains the same while the interpretation changes homomorphically. This identification of type terms and function terms, first formalized in $\lambda$-typed $\lambda$-calculi, has been applied in Deva. Moreover, since the latter is essentially a programming language, its design, implementation, and use benefit from well-established methods: classical composition operators are introduced, operational semantics serves as a formal definition, implementation techniques are available, and teaching material as well as support tools follow standard principles. The difference with ordinary languages is, of course, the application domain: the types serve here to express propositions such as specifications or programs, rather than just data classes.

Model case studies played an important part in design. This has been one of the benefits of continued cooperation with good industrial partners. A primary objective was to formalize effective methods of software design. In industry, the most productive methods use successive refinements from state-based specifications; the first example in the book illustrates this approach for an application in the field of biology. A promising research direction is the derivation of efficient programs on the basis of algorithm calculi; this is presented here in another case study. Such experiments and existing models of enlightening proofs have continuously influenced the design of Deva. In consequence, its description has been significantly modified a number of times. The genericity and reusability of implementation tools helped in mastering this necessary evolution. In fact, the current version of Deva may well be adapted further.

Before formalizing a topic, we must first understand it and design a good theory for it. In the case of program derivations, this theory-building comprises three layers: there are basic theories from mathematics and computing, then theories of design methods and application domains, and finally theories for specific program derivations. A significant part of formal software development is thus concerned with classical mathematics. This appealing blend of mathematics and programming science could be termed *modern applied mathematics*. The elaboration of a theory must not be confused with its formalization. On the one hand, without an adequate theory, the formalization does more harm than good: the better a theory is, the happier its formalization. On the other hand, one should be able to take any good piece of mathematics and formalize it nicely in a proposed language for *formal enlightening proofs*. Once a design method has been given a good theory and has been formalized accordingly, it is possible to develop formal proofs of theorems about the method itself. The present book, for instance, provides a formal theory of reification, and then a formal proof of the transitivity of reification.

Various languages for formal enlightening proofs are currently being experimented with. The reprogramming of common theories in these languages appears to be counter-productive; it is reminiscent of the republication of similar material in several books or the recoding of software libraries in different programming languages. Happily, the cost of repeated formalizations can be reduced: where the languages are quite different, at least the contents of the theories can be communicated using literate formalization, as in the case of the present book; if the languages are similar, specific translators can be developed to automate recoding. The latter solution can be used in the case of successive Deva versions.

The following views may underlie further work. Firstly, to the extent that proof expressions are homomorphic to proposition expressions, we must be free to work at the level of functions or at that of types. This would allow us to express a proof step not only as the application of a function typed by a rule to a constant typed by a proposition, but also as the direct application of the rule to the proposition; this better matches the nature of enlightening proofs.

Secondly, purely linguistic aspects play a major part: compositional structures, formal beauty, and stylized notations prove crucial for successful intellectual communication. Thirdly, semantics must be understood by minds and not just by machines: to foster higher-level reasoning on proof schemes, algebraic laws are more useful than reduction rules. The formalization of enlightening proofs should add neither semantical nor syntactical difficulties: it must instead clarify the proofs even better. Fourthly, it should be possible to formalize well any component of mathematics: the scientific basis for the design of software systems tends to include any mathematics of interest in system design. Finally, it is mandatory to capitalize on existing symbolic algorithms, decision procedures, and proof schemas; ideally, these should be integrated in specific libraries so as to be understood, communicated, and applied. In a word, we must apply, in the design and use of high-level proof languages, the successful principles established for existing high-level programming languages. The correspondence between proofs and programs also results from the similarities between algebras of proofs and algebras of programs, not only from the embedding of programs within proofs.

To conclude, Deva can be seen as a tentative step towards a satisfactory language of formal enlightening proofs. The authors should be warmly thanked for presenting this scientific work to the computing community.

Michel Sintzoff

# Acknowledgments

This book would not have been possible without the continuing support of several people. First of all, we wish to thank Michel Sintzoff, who is, so to speak, the father of the research reported on in this book. His ideas on formal program development were the starting-point for the development of Deva. We are grateful to him for the stimulating thoughts he shared with us, his constant supervision, and his kind and inspiring overall support. We hope that he still recognizes some of his ideas when reading the following pages. Next, we wish to thank Stefan Jähnichen for his guidance and support. His interest in turning theoretical results into practical applications and his insistence on testing Deva on non-trivial case studies greatly influenced our work. Our thanks also go to the other members of the TOOLUSE project, who patiently experimented with the language and who bore with us, in the early stages, during the frequent changes to the language definition. Their comments and suggestions greatly contributed to the design of Deva. We acknowledge in particular the contribution of Philippe de Groote, who invented and investigated a kernel calculus of Deva. Jacques Cazin and Pierre Michel closely followed the design of Deva and gave helpful comments and advice. Pierre-Yves Schobbens shared with us his knowledge of VDM. We also thank the UCL at Louvain, the Belgian project Leibniz (contract SPPS/RFO/IA/15) and the GMD at Karlsruhe for their financial support beyond the duration of the TOOLUSE project. The staff of both the UCL and the GMD were also very supportive and took great interest in our work. We wish to express our gratitude to all of these people for their encouragement and friendship. It was this that made the whole project an enjoyable and worthwhile experience.

We had an equally stimulating working environment in Berlin. The BKA group, in particular, provided inspiring insights. Martin Beyer and Thomas Santen read drafts of the book and made valuable comments. Two implementation efforts which are underway in Berlin have had a significant impact on the book. First of all, there is Devil (Deva's interactive laboratory), an interactive environment for developing Deva formalizations, which is being designed and implemented by Matthias Anlauff (aka Maffy). We used Devil to check (almost) all Deva texts contained in this book. Our very special thanks to Maffy for his tremendous implementation effort. He worked night and day incorporating new features to extend the power and usability of the system, and removed the few bugs we discovered. Secondly, there is the DVWEB system, a WEB for Deva, which is being implemented by Maya Biersack and Robert Raschke, and which we used to write the whole book. On the one hand, DVWEB enabled us to work on a single document for both Devil and TeX, and on the other hand, its macro features greatly improved the presentation of Deva formalization. Our thanks also go to Maya and Robert for this valuable tool. Furthermore, we wish to express our gratitude to Phil Bacon for polishing up the final text.

Finally, we wish to express our gratitude to the many known and unknown referees for their helpful criticism and advice. Professor Goos, in particular, helped us to state our objectives more clearly by providing a number of critical comments on the initial draft of the book.

# Table of Contents

# 1 Introduction

The present book presents *Deva*, a language designed to express formal development of software. This language is generic in the sense that it does not dictate a fixed style of development, but instead provides mechanisms for its instantiation by various development methods. By describing in detail two extensive and quite different case studies, we document the applicability of Deva to a wide range of problems.

Over the past few years, the interest in formal methods has steadily grown. Various conferences, workshops, and seminars on this topic have been organized and even the traditional software engineering conferences have established their own formal method sessions. A journal devoted entirely to this subject, entitled "Formal Aspects of Computing", has also been started. The association "Formal Methods Europe" was founded in 1991, as a successor to "VDM Europe" to promote the use of formal methods in general by coordinating research activities, organizing conferences, etc. Formal methods can thus no longer be viewed as the exclusive reserve of theoreticians.

However, despite the fact that current research is concerned with the whole range of activities involved in the software development process, the industrial application of formal methods is mostly limited to specification. Languages such as VDM or Z are enjoying growing acceptance in industry, as is evidenced, for example, by a number of articles in the September 1990 issue of "IEEE Transactions on Software Engineering", two special issues of "The Computer Journal" (October and December 1992) on formal methods , or by ongoing efforts to standardize both languages. This success is due to the fact that a formal specification allows formal reasoning about the specified system; in other words, it enables the question as to whether some desired property is true for the system to be answered by a mathematical proof. This greatly reduces the risk of errors that would be detected at a much later stage in the development process or not at all, thus justifying the allocation of more time and resources to the specification phase. On the other hand, truly formal methods are rarely used beyond the specification phase, i.e., during actual development. Even in pure research environments, completely formal developments remain the exception.

But what exactly *is* a formal method? According to Webster's Collegiate Dictionary, a method is "a procedure or process for attaining an object", this object being, in our case, a piece of software. A method is called *formal* if it has a sound mathematical basis. This means essentially that it is based on a formal system: a formal language, with precise syntax and semantics, some theories about the underlying structures (theories of data types, theories of refinement, etc.), and a logical calculus which allows reasoning about the objects of the language.

The main activity performed during the specification phase is the modeling of the problem in terms of the language. During the development phase, it is the refinement of the specification down to efficient code using techniques such as data refinement, operation decomposition, or transformations. This process is, again, expressed in terms of the language, and the proof obligations associated with each refinement step are carried out within the logical calculus making use of the method's theories.

The major reason why formal developments are so rare is that far more proofs are required during development than during the specification phase. In fact, the nature of the proofs called for during the specification phase is quite different from that of the development phase. During specification one usually proves properties that are desired to hold true for the specified model. This is done in order to convince oneself or the customer of the adequacy of the specification. During design, however, one is obliged to discharge all the proof obligations associated with a refinement or transformation step, so that, in the end, one can be sure that the product of the design is correct with respect to the specification. These proof obligations are, in most cases, not profound, but fairly technical. It frequently happens that a proof obligation which is obvious to the developer requires a tricky and lengthy proof. Generally speaking, the amount of work involved in discharging a particular proof obligation is quite disproportionate to the quality of the new insights gained into the product.

The burdensome requirement of proving every little detail of a proof obligation is therefore relaxed by most methods to the point where the proof obligations are stated in full without the need to prove all of them. We call methods which adhere to this paradigm *rigorous*. Although, with a rigorous development, it is once again up to the designer to decide whether he is satisfied that the result meets the specification, it is, in principle, still possible to prove the development correct. (One might object that this is similar to the situation faced when verifying a piece of code, but the crucial difference is that, during the development process, all the vital design decisions have been recorded together with the proven and unproven proof obligations.)

Here, it may be asked why it is not possible to give reasons for the correctness of a development in the same way a mathematician gives reasons for the correctness of a proof in the first place. In fact, there is no proof given in any mathematical text we know of which is formal in the literal sense. Instead, proofs are presented in an informal, descriptive style, conveying all the information (the exact amount depends on how much background knowledge is expected of the reader) necessary to construct a formal proof. However, there are a vast number of proofs to be carried out during development, and the traditional mathematical procedure of judging proofs to be correct by submitting them to the mathematical community for scrutiny is inadequate in this situation. It must also be remembered that, here, for the first time, an engineering discipline is faced with the task of producing, understanding, and managing a vast number of proofs — a task whose intellectual difficulty is not to be underestimated. Machine support is therefore needed and this calls for formality. In this sense, we call a proof *formal* if its correctness can be checked by a machine.

However, despite the fact that full formality is needed to enable a machine ultimately to check a proof, this cannot mean that one is forced to give every little detail of a proof. This would definitely prevent formal developments from ever gaining widespread acceptance. The aim should be to come as close as possible to the informal way of reasoning (from the point of view of the designer), while at the same time remaining completely formal (from the point of view of the machine).

In this book, we present a *generic development language* called *Deva* which was designed to express formal developments. Syntax and static semantics can be checked by a machine, their correctness guaranteeing the correctness of the development. In order to ensure independence from a specific development methodology, a major concern during design of the language was to isolate those mechanisms essential for expressing developments. Accordingly, the language allows us to construct from these basic mechanisms new mechanisms specific to a particular method. In this sense, Deva may be said to be a *generic* development language, since the formal language underlying a specific formal method can be expressed in terms of Deva.

## Ideal requirements for generic development languages

The above discussion yields in several ideal requirements for generic development languages which we now go on to summarize. We will subsequently show how and by what means Deva satisfies these requirements.

First of all, a development language must provide a medium for talking about specifications, developments (i.e., refinements, transformations, and proofs), and programs. A generic development language must, in addition, provide means for expressing mechanisms for the developments themselves.

Good notation is a frequently neglected aspect of languages, and yet it is one of the most important as regards usability and acceptance [6]. Good notation should be as concise as possible, but, at the same time, suggestive enough to convey its intended meaning. In the context of development languages, this means that the notation should support various different ways of reasoning and development styles, the notational overhead introduced by formality being kept as low as possible. The developments expressed in this formal notation should compare favorably in style and size with those demanded by rigorous methods. A generic development language must, in addition, provide means for defining a new notation in a flexible and unrestrictive manner.

The language must provide means for structuring formalizations. The lesson learned from programming languages is that structural mechanisms are indispensable for formalization, even on a small scale. In the context of generic development languages, this is even more important because of the wide range of different levels of discourse. Hence, such a language must provide mechanisms for structuring in-the-large and in the small. For structuring in the large, this means that the language must have some sort of modularity discipline, which includes definition of modules, parameterizations and instantiation of modules, and inheritance among modules. Experiments have shown that, for formal developments, the following mechanisms are useful for structuring in the small: serial and collateral composition of deductions, declaration of axioms, abbreviation of deductions, parameterization and instantiation of deductions.

Since proving is one of the most important activities in formal development, it must be possible to express proofs in the language. The correctness of such proofs must be checkable with respect to an underlying deduction system. Since a formal method usually comes with its own deduction system and underlying logic, a generic development language must be flexible enough to handle a variety of logics and deduction systems.

As we have argued above, the amount of detail required to enable a machine to check the correctness of a proof has a considerable influence on the usability of the formal approach to software development. Ideally, we envisage a situation where the designer gives a sketch of a proof, just as a mathematician would, and lets the machine fill in the details. But, this is not yet state-of-the-art, and so we must be a little more modest in our demands. The language should, however, go as far as possible in allowing incomplete proofs and should also incorporate some basic mechanisms for defining so-called tactics — which can be understood as means for expressing proof sketches. Functional languages such as ML have been used with considerable success to program recurring patterns of proof into tactics and to design systems supporting semi-automatic proofs based on tactics. It is certainly a desirable goal to completely automate the task of proof construction. So far however, this approach has been successful only in very limited areas, and, all too often, has resulted in systems that obscure rather than clarify the structure of proofs.

Of course, the language should be sound in the sense that any errors contained in a formalization must be due to the formalization itself and not to the language. For example, a correct proof of a faulty proposition must ultimately result from the (correct) use of a faulty axiom in one of the underlying calculi rather than from an internal inconsistency in the language.

Finally, the language should be supported by various tools. The most important tool is certainly a checker, which checks the correctness of a formalization expressed in the language. Around such a checker, a basic set of support tools should be available. Users should be able to experiment with their formalizations in an interactive environment; the user should be able to draw on a predefined set of standard theories containing formalizations of various logics, data types, etc; likewise, they should be able to store their formalizations for later reuse; and they should be assisted in preparing written documents containing formalizations.

Note that all the above requirements for a generic development language are intended to guarantee that one can express, or better formalize, the formal system underlying a formal method. We do not intend to deal with other aspects of a method such as recommendations, guidelines, and heuristics. Thus, when we speak, in the sequel, of formalizing a method, we invariably mean the formalization of the underlying formal system. This implies that formalization of a formal method in terms of a generic development language gives no indication as to how to *invent* a development; this is left to the pragmatics of the method and to the intelligence of the designer.

# The Deva Approach

We now wish to describe the concrete approach adopted when developing Deva with a view to meeting the above requirements.

The most important design decision was the choice of a higher-order typed $\lambda$-calculus as a basis for the language. This decision was motivated by several considerations. Typed $\lambda$-calculi have served as the basic formalism for research into the formalization of mathematics. The languages which grew out of this research include, for example, the AUTOMATH family of languages [27], [28], [29], [80], the Calculus of Constructions [24], the Edinburgh Logical Framework [50], and the NuPRL system [22] [70]. These so-called Logical Frameworks were mainly used for formalization of mathematics, functional programming, and program synthesis from proof (cf. [54] and [53]). One of the main results of this research was that these logical frameworks proved to be an effective approach to the task of formalization in general, and one which is also amenable to implementation on a machine. The underlying principle here is the so-called Curry-Howard paradigm of 'propositions-as-types' and 'proofs-as-objects'. We do not wish to go into greater detail at this point, but the basic idea is that there is a one-to-one correspondence between the propositions of (constructive) logic and the types of a typed $\lambda$-calculus, and between the (constructive) proofs of propositions and the objects of a type. Given this correspondence, proving amounts to functional programming, and proof-checking to type-checking, which is what makes the approach so attractive for implementation on a machine. In the next chapter, we explain this paradigm in more detail and present a number of intuitive examples.

Starting from this design decision, we proceed as follows: we wish to view specifications, programs, and deductions as formal objects which can be formally manipulated and reasoned about, and for which we can formulate correctness properties. For specifications and programs, this is nothing new — they are considered to be formal objects of study in other contexts as well. However, in the case of deduction, it is a somewhat new perspective: deductions are viewed as formal objects relating specifications to programs. This is the key concept in our approach to the formalization of formal methods. When formalizing a method, we do so by stating axiomatically, among other things, which deductions are allowed by the method. Such an axiom describes how a specification is related to a program by a particular deduction.

In the context of a typed $\lambda$-calculus, we realize this aim by representing specifications, programs, and deductions as $\lambda$-terms. They can be manipulated and reasoned about with the usual machinery that comes with such a calculus. Correctness and consistency issues are handled by the typing discipline of the calculus.

A particularly well-suited logical framework was selected as the starting point for the design of the Deva language: Nederpelt's $\Lambda$, one of the latest members of the AUTOMATH family [80]; see [33] for a recent presentation of $\Lambda$ in the spirit of Barendregt's Pure Type Systems [10]. This calculus was chosen, after a number of others had been evaluated, because it is comparatively simple and economical, and because it supports some of the major concepts of structuring in the small, namely parameterization and instantiation of deductions. Although $\Lambda$ has not been experimented with in the AUTOMATH project, it does constitute a major scientific contribution of that project. However, $\Lambda$ remains very much like an ordinary $\lambda$-calculus: it is based primarily on binding and substitution, and it fails to provide composition, product, and modules, such as are needed in our approach to program development. The definition of Deva grafts these concepts on to $\Lambda$. A second major extension to $\Lambda$ concerns the distinction, in Deva, between an explicit and an implicit level. The explicit level includes all the extensions to $\Lambda$ we have just mentioned. The implicit level adds constructs which are instrumental in meeting another important requirement: that of allowing incomplete sketches of deductions, proofs, etc. Parallel to these extensions to $\Lambda$, the normalization proofs established for $\Lambda$ were adapted. These language-theoretical properties are important for demonstrating the soundness of the language, i.e., that Deva itself does not introduce errors into a formalization. To summarize: Deva is to $\Lambda$ what a functional language is to the pure $\lambda$-calculus.

## Tool Support

Right from the beginning of the design process, prototypical implementations of type checkers and other support tools for Deva were built and experimented with. These prototypes were not, however, intended for use in full-scale Deva developments, but were rather developed for experimentals purposes. Hence, they supported only selected features of the language's functionality as presented in this book. The two case studies examined here provide ample evidence that medium-scale formal developments are feasible, provided that the user is assisted by the machine via a set of tools.

The design and implementation of such a tools for the full Deva version is the subject of a current Ph.D. thesis [3]. Initial (beta-) versions of this tool-set — called "Deva's Interactive Laboratory" or "Devil" — have been available since late 1992, and they are currently being used for a number of ongoing case studies (e.g. [12], [89], see below). Since the tools are being continously further developed, we give only a brief summary. The structure of the tool-set is shown in Fig.1. This diagram illustrates the current state of the support environment. Direct user experiences will shape its future development. A syntax-check, a consistency-check and an explanation (cf. Chap. 3) constitute the central components of the system. Once a formalization has been checked, it can be stored in an efficient ("compiled") form for later retrieval. The interactive design of and experimentation with Deva formalizations is made possible through an interactive user interface for which both a plain TTY and an X-Windows-based realization exist. Through a database, the user may access previously defined or compiled formalizations.

The design of formal specifications, formal developments or any other formalizations should go hand in hand with the design of their documentation. In fact, good documentation of formal specifications or developments is even more important than documenting or "commenting" programs, because, like any other

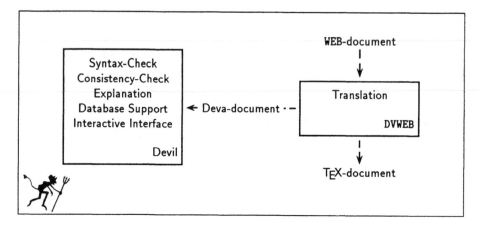

**Fig. 1.** Structure of the support environment

specification or development, formal specifications and developments are primarily intended as a means of *communicating* with other people. Knuth gives the following motivation for *literate programming* in [63]: "Instead of imagining that our main task is to instruct a *computer* what to do, let us concentrate rather on explaining to *human beings* what we want a computer to do." A similar statement might be applied to formal specifications or formal developments: Instead of imagining that our main task is to explain to a computer why a specification or development is correct, let us concentrate rather on explaining to human beings why they are correct. To realize his idea of literate programming, Knuth designed the WEB system of structured documentation. Originally intended for Pascal programs only, various WEB's are being developed for other languages and formalisms (cf. Knuth's recent book on literate programming [64]). Such a WEB tool is also being developed for Deva. The user writes a single WEB document which combines the documentation and the formal Deva code. The presentation of the formalization can be given in a natural, *web*-like, manner, unhindered by the syntactic restrictions of the formal language (an example being this book!). The Deva code portions are pasted together in a preprocessing step to produce a formal Deva document which can then be checked by the Devil system. A second preprocessing step produces a TEX-document which maintains the web-like structure of the presentation and in which the Deva code portions are typeset in an esthetical manner. Furthermore, an index of variables is produced. Details of DVWEB are given in [13].

Since all of these tools were not available at the time we started writing the book, all the formalizations contained in this book were just typed in. The only "tool" around was a set of LATEX-macros which made this task a little less painful. But with these tools available, we decided to rewrite all the chapters containing Deva formalizations, so that Chaps. 2, 4, 5, and 6 are now self-contained WEB-documents which have been checked by the system. It is worth noting that the Devil system revealed numerous errors in the original document. While most of these errors were easy to correct, some pointed to serious flaws in the reasoning and required significant repair efforts.

## Intended Readership

The above introduction and overview of the language should have made it clear — and we will try to demonstrate this in the course of the book — that generic development languages in general, and Deva in particular, offer a useful framework for tackling formal developments. Four different application areas for Deva come to mind. Deva can be used to formalize a method in order to

- impart a precise understanding of the method,
- spot shortcomings in the method or its documentation,
- experiment with libraries of basic theories for software developments based on the method, or
- obtain a prototypical support environment for the method so that formal developments may be constructed, documented, and checked for correctness with respect to the formalization.

The book can be read by anyone with a basic background in formal approaches to software development, as, for example, is given in [60].

## Synopsis

The book is divided into two parts. The first part describes the Deva language; in the second documents two case studies. Part one comprises

- Chapter 2 which gives an introduction to Deva, presenting intuitive examples of theories and proofs chosen from elementary mathematics and logic. The goal is to convey an intuitive understanding of the use and properties of the language.
- Chapter 3 which presents and explains the formal definition of Deva and briefly summarizes some theoretical results and still open questions. The goal is to convey a thorough technical understanding of the notation and its design. The material contained in this chapter is based on [104] and [105]. On a first reading, this chapter may be skipped.

Part two comprises

- Chapter 4 which presents a selection of basic logical and mathematical theories, formalized in Deva. The goal is to demonstrate the principal formalization power of Deva on a number of well-known examples.
- Chapter 5 which presents a case study on the formalization of VDM-style developments in Deva. In particular, VDM data reification is formalized. Then, a data reification step from a VDM development in the context of a biological case study is formalized in detail. The material presented in this chapter is based on [67] and [103]. Finally, the formalized data reification is formally proven to be transitive.

– Chapter 6 which presents a case study on the formalization of algorithm calculation. It consists of a formalization of representative parts of Squiggol, also known as the "Bird-Meertens formalism", an algorithmic calculus for developing programs from specifications. A selection of the structures and laws of Squiggol and two complete program developments are described in Deva. The material presented in this chapter is based on [102].

The two case studies can be read independently of each other. It is no accident that they deal with quite different areas; in fact, they have been selected to illustrate the genericity of Deva and to address a wider readership.

## Historical Background

The design of an early precursor of Deva was set out by Michel Sintzoff in a series of papers ([91], [92], and [93]) in the early '80s. Deva itself, as presented in this book, was developed mainly between 1987 and 1989 in the context of the ESPRIT project TOOLUSE ([57], [20], [95]). The objective of the TOOLUSE project was to study a broad spectrum of development methods (e.g. Jackson System Design, VDM, or Burstall-Darlington's fold/unfold method for program transformation) and to design a method-driven support environment. Deva was intended to serve as a notational framework to help promote an understanding of such methods. The language was developed by the collaborative effort of three different project subgroups: a group headed by Michel Sintzoff at the Université Catholique de Louvain (UCL), a group headed by René Jacquart at the French Research Center for Technology (CERT) in Toulouse, and a group headed by Stefan Jähnichen at the German National Research Center for Computer Science (GMD) in Karlsruhe. A kernel calculus of Deva was proposed by Philippe de Groote, who also developed its language theory [30], [31], [33], [32]. In the course of the project, several prototype support tools for the evolving versions of Deva have been developed [41]. This book presents the complete language as set out by one of the authors in [104].

## Related Approaches

Before turning our attention to currently evolving approaches, we would like to mention a pioneering experiment in the formalization of mathematics conducted in the context of the AUTOMATH project mentioned above. It consisted of the translation of Landau's "Grundlagen der Analysis" into one of the AUTOMATH languages, a translation that was completely checked by machine [99].

Machine support for formal development can be roughly divided into specific support, i.e., support for a single and fixed logic or programming method, and generic support, i.e., support for a range of logics or methods. While the focus of this discussion will be on generic systems, we wish to mention first of all several currently evolving systems that demonstrate the usability of two key techniques for making formal program developments more accessible to humans: The first technique is to program recurring patterns of proof into *tactics* [44]; the second is to express domain knowledge in abstract development schemes. The *Karlsruhe Interactive Verifier* KIV [51] uses tactics to implement high-level strategies for the development of verified imperative programs. Similarly, in the *Larch Prover* LP [42], a variety of tactical mechanisms are used to provide interactive proof support. LP has been experimented with in a variety of case studies, including the debugging of module interface specifications [48]. The *Kestrel Interactive Development System* KIDS [96] uses a hierarchy of algorithm design schemes such as divide-and-conquer and dynamic programming for schema-driven interactive development.

Of the generic systems, some are primarily oriented towards formalization of logics, and others towards program developments, which means a further sub-division. In fact, a number of generic systems of both kinds are under development, some having been just recently announced, so that it becomes rather difficult to discuss all of them in detail. Instead, we wish to draw attention to three specific generic approaches. All three approaches are characterized by an early focus on building or improving an interactive support environment for — a significant portion of — formal proofs. The design of Deva, which was begun at a later date, is characterized rather by the successive approximation of a notation which could express some representative styles of formal program development reasonably well. Experiments and comparisons were conducted in paper-and-pencil form, using quickly constructed Deva prototypes, or with existing systems from related approaches. This may help to explain some of the differing design choices.

The **B**-Tool is a rule-based inference engine with rule-rewriting and pattern-matching facilities [1]. Initially, the **B**-Tool concentrated rather on automatic proving; explicit proof mechanisms and tactics were added later. The **B**-Tool is generic in the sense that it has no pre-defined encoding of any specific logic. It can be configured to support a variety of different logics by specifying them as so-called "theories". To describe these logics, the **B**-Tool offers a number of built-in proof mechanisms, such as reasoning-under-hypotheses, scoping-for-variables, a notion of quantifiers, metavariables, substitution, equality, etc. Proof strategies may be described in the **B**-Tool by tactics. Both forward reasoning and backward reasoning are supported. The **B**-Tool has been used and tested in a number of formal program developments, each of them typically requiring several hundred mathematical theorems to be proved.

The Isabelle system is an interactive theorem-prover based on intuitionistic higher-order logic [83]. It allows support of the proof construction in a variety of logics. Its main orientation is towards the machine support of proof synthesis. To this end, it offers powerful proof tactics and a concept of backwards proof construction. Unlike Deva, Isabelle is not based on propositions-as-types, but uses instead predicates to formalize the propositions and theorems of various logics. This entails a number of — relatively small — technical complications in the formalization of logics. Isabelle benefits from the firm foundations of intuitionistic higher-order logic, which allow adequacy of formalizations to be proven, and the reuse of well-known techniques such as higher-order unification. A recent article describes the experimentation with Isabelle to prove theorem about functions in Zermelo-Fraenkel set theory [81]. The overall orientation towards genericity and sound foundations makes Isabelle similar to Deva. However, the two approaches do quite differ in practice, perhaps because Deva is oriented more towards theories and program development, while Isabelle is oriented more towards interactive proof synthesis. In addition to Isabelle, several similar systems based on higher-order logic are being developed, some of them oriented towards hardware design [4]. The most widely used one among these is probably the HOL system [45] [46]. A translation of dependent type theory into HOL is proposed in [56].

The mural system is a formal development support system consisting of a generic reasoning environment and a support facility for VDM [61]. The reasoning component is based on a logic with dependent types. mural can be instantiated by a variety of logics. Logical theories are organized in a hierarchical store, containing declarations, axioms, derived rules, and proofs. The main emphasis in the design of mural has been on the interface for interactive proof construction and theory organization. Proofs are constructed interactively in a natural deduction style. The VDM support component provides support for the construction of VDM specifications and refinements. It also generates proof obligations stating the correctness of refinements. The proofs themselves must then be constructed inside the reasoning environment. In [39], the application of mural is demonstrated in the the specification and verification of a small VDM development. The mural system could easily be extended by support facilities for methodologies other than VDM. The Deva approach shares with mural the orientation towards genericity and method support. Compared with the B-Tool, Deva was geared to a more powerful use of logical genericity. For example, the formalization of VDM presented in this book can be viewed as a *formal specification* of selected parts of VDM. This VDM specification is then used in the book not only to reason *inside* VDM but also to reason *about* VDM, by proving the transitivity of VDM data reification.

Furthermore, there are a number of relatively new generic systems implementing generalized typed $\lambda$-calculi; these include LEGO [74], Coq [36], and ALF [75]. An application of the LEGO system to formalize various logics and proofs is described in [5]. LEGO has also been used to formalize program specification and data refinement in the extended calculus of constructions [73]. For an application of the ALF system, we refer to [23]. Also, we would like to mention the logic programming language Elf [85], based on the Edinburgh Logical Framework [50]; Elf can be used to encode a dependent-type $\lambda$-calculus [38] or to directly formalize theories and proofs [86].

## 2 Informal Introduction to Deva

In this chapter we will give a tutorial-like introduction to the Deva language. Our aim is twofold: On the one hand, we want to introduce the most important elements of the language and motivate their use. On the other hand, we want to illustrate some techniques which have proved useful during the various case studies. We begin with introducing the central elements of Deva in the setting of a complete, self-contained example. Next, we show how different proof styles can be expressed in Deva. Finally, some advanced constructs of Deva are described.

### 2.1 An Algebraic Derivation

The guiding example will be a *completely* formalized proof of the binomial formula $(a + b)^2 = a^2 + 2ab + b^2$. This example might strike the reader as a somewhat odd choice for the introduction of Deva. However, as we have argued in the introduction, proving and deducing are central activities of formal software development. In particular, proofs of algebraic properties occur frequently, such as, for instance, program transformation laws. Therefore, it makes sense to choose the formalization of a proof as an introductory example. We chose this particular example because it lends itself easily to a complete and reasonably compact formalization. The formalization of any sensible program development would have involved considerably more detail, only distracting from the purpose of this introduction, namely to introduce the language Deva. The two case studies presented in this book will illustrate the complete formalization of central aspects of formal program development methods.

#### 2.1.1 The Problem

Suppose we ask a mathematician to prove that the binomial formula

$$(a + b)^2 = a^2 + 2ab + b^2$$

is valid for the natural numbers. She would probably reply "That's obvious, Q.E.D! Don't bother me with such trivial stuff." So let us further suppose that we dare to bother her a bit more and ask for a *formal* proof. If she is in a good mood, she will probably produce something similar to the following derivation:

$$(a + b)^2$$
$$= \quad \{ \text{ definition of squaring } \}$$
$$(a + b) \times (a + b)$$
$$= \quad \{ \text{ distributivity } \}$$
$$a \times a + a \times b + b \times a + b \times b$$
$$= \quad \{ \text{ commutativity } \}$$
$$a \times a + a \times b + a \times b + b \times b$$
$$= \quad \{ \text{ definition of squaring, definition of "doubling" } \}$$
$$a^2 + 2 \times a \times b + b^2$$

Since we do not want to get her really upset we are satisfied with this proof. But only a little thought reveals that this proof is not really a formal proof but rather a rigorous one. Too many assumptions are left implicit such as axioms and properties of the natural numbers which the proof refers to (for instance, what does "squaring" mean?), or the formal inference system in which the proof is carried out. All these implicit assumptions have to be stated in detail if the proof has to pass a thorough examination, for example by a machine. In the following sections of this chapter we will present a complete formalization of the proof of the binomial formula within Deva. The result is shown in Fig. 6.

### 2.1.2   The Structure of the Formalization

It was said in the previous section that a complete formalization of the proof of the binomial formula must address all aspects of the proof. In fact, the formalization we will present, consists of a part formalizing basic properties of a parametric equality relation, a part formalizing the natural numbers, and a part formalizing the proof itself. Deva distinguishes two syntactic classes which are called *contexts* and *texts*. In general, contexts are used to *structure* theories such as algebraic theories, logical theories, or development methods. Texts on the other hand serve to *express* objects of theories such as axioms, formulas, proofs, or program developments. Therefore, we will structure our formalization by means of contexts. The following context illustrates the global structure of the formalization:

**context**  *BinomialFormulaDevelopment* :=

$[\![ \langle$ Preliminaries. 2.1.3 $\rangle$

; $\langle$ Parametric equality. 2.1.4.5 $\rangle$

; $\langle$ Natural numbers. 2.1.5 $\rangle$

$]\!]$

Some simple elements of the Deva language are used by the preceding context: The syntax of Deva defines a context to be a sequence of *definitions* and *declarations* separated by semicolons and bracketed by the context brackets '$[\![$' and '$]\!]$'. In case of a single definition or declaration, the brackets can be omitted. The preceding context introduces one particular form of definition, namely the *context definition* which associates a context with an identifier and is indicated by the keyword **context** . Thus, we have a single context which is a definition introducing the name *BinomialFormulaDevelopment*. Later, we will see how a context named by an identifier is *used* within another context.

The notation "$\langle$ Preliminaries. 2.1.3 $\rangle$" stems from the DVWEB system (cf. the introduction). It stands for the Deva code defined in Sect. 2.1.3 on page 16. During the process which takes a WEB-document and translates it to the corresponding Deva-document these references are resolved by textually replacing them with their definitions. The reader may read a "WEB" in a similar style: When encountering such a reference, one looks up the corresponding definition.

The section number makes it easy to locate the definition. In general, such a definition begins with "⟨Name of the section. Sect.No.⟩ ≡" and is followed by its replacement Deva code. A short note below the code tells the reader in which section(s) the defined code is used. The appendices contain tables of all these "named Deva code sections" in addition to an index of the variables used. Sometimes, the definition of a replacement text may stretch over several sections. In that case, the code sections following the first one begin with "⟨Name of the section. Sect.No.⟩ +≡" where the name is always the same. A short footnote tells the reader in which other sections the definition of the replacement text is continued. The complete replacement text is then the concatenation of the single replacement parts in the order they appear in the text.

Returning to our formalization example, we now ask how we formalize the different parts? We mentioned above that *texts* make up the second syntactic class of Deva. This fragment of Deva is a variant of a *typed λ-calculus with dependent types*, enriched by several operators useful for expressing formal proofs or developments. Hence Deva is a member of the family of *logical frameworks* whose basis is some variant of a higher-order λ-calculus. Representatives of such frameworks were mentioned in the introduction and it was also said that they are based on the so called Curry-Howard paradigm of "propositions-as-types" and "proofs-as-objects". According to this paradigm, to prove a proposition, one constructs an element of a type which is the interpretation of the proposition. For example, in the particular case of the binomial formula, we will eventually (in Sect. 2.1.6) define the Deva-text *thesis*

$$
\begin{array}{l}
[\ \ a, b : nat \\
\vdash (a + b)^2 = a^2 + 2 \times a \times b + b^2 \\
]
\end{array}
$$

which we regard as the formalized version of the binomial formula. Then, we construct a Deva-text whose type is exactly this text and which we regard as a proof of the binomial formula by the Curry-Howard isomorphism.

In the introduction we said that we want to treat deductions or program developments as formal objects. To be more precise, we will consider program developments to be proofs. According to the Curry-Howard isomorphism, the type of a program development is therefore a proposition. For example, in the framework of the Bird-Meertens formalism, the type of a program development takes the form $P = Q$ where $P$ and $Q$ are functional expressions of some sort. Usually, $Q$ has a more efficient interpretation as a functional program than $P$. A proof of $P = Q$ may then be regarded as a development of the program $Q$ from the specification $P$. Consider a second example: In the context of a data-reification technique, the type of a program development may take the form $V_2 \sqsubseteq V_1$ where $V_1$ and $V_2$ are two versions of a system specification. The symbol $\sqsubseteq$ denotes that $V_2$ is using "more concrete" data-structures than $V_1$. A proof of $V_2 \sqsubseteq V_1$ may then be regarded as a verified data-refinement step. Finally, consider a third example: In the context of imperative program verification, the type of a program development may be an assertion triple $\{P\}prog\{Q\}$ where $P$ and $Q$ denote pre- and postcondition of the program *prog*. A proof of

$\{P\}prog\{Q\}$ may then be regarded as the verification of program *prog* wrt. $P$ and $Q$.

However, the last example is somewhat problematic: We do not want to hide the fact that imperative program logics pose some problems when trying to make a natural formalization in Deva. The general problem is that the global state, or store, must be made explicit. Technically, this is not a very difficult exercise. However, since Deva is a language of functional nature, all obvious solutions result in a significant change of the appearance and the use of the assertion logic. This can be considered as a drawback, but these issues are not a topic of this book. The interested reader is referred to [104] for a formalization of a simple logic of assertions in Deva.

Let us continue with our example. The next few sections will be concerned with the formalization of the preliminaries, the parametric equality, and the natural numbers as outlined above. We will introduce new language constructs and techniques for formalizing if and when the need arises. In this way, we will be a bit informal and we often will not tell the whole story. But we hope that the reader who is not already familiar with the Deva language and this approach towards formalization will gain an intuitive understanding in the course of this chapter. A formal treatment of the language will be given in the next chapter and it might be a good idea to consult it once in a while in order not to lose track of syntax and semantics.

### 2.1.3   Preliminaries

A formalization of the proof of the binomial formula will essentially have to specify two things: the natural numbers and a logic which permits to express the proof. Looking at the proof given in Sect. 2.1.1 one recognizes that a theory of equations will suffice as the underlying logic. Such a theory may be expressed in terms of a parametric sort, independent of a theory of natural numbers. We will therefore assume two primitive types, sorts and propositions, and this expressed in Deva by the following context:

$\langle$ Preliminaries. 2.1.3 $\rangle \equiv$

$[\![$ *sort* : **prim**

; *prop* : **prim**

$]\!]$

This code is used in section 2.1.2.

The context shown above introduces several new elements of the Deva language. First, the context consists of two *text declarations*. Similar to a context definition, a text declaration is an atomic context. The first text declaration declares the *text-variable sort* to be of *type* **prim** and the second one declares the text-variable *prop* to be also of type **prim** . In Deva, a text has a type which is again a text. This is contrary to a conventional higher-order $\lambda$-calculus where there is a syntactic difference between the object level and the type level. However, we argue that these two levels have identical structure and that this observation leads to a more homogeneous language design. The basic building block for constructing texts is the *text constant* **prim** . It is itself not typeable in order to prevent a text from being typeable infinitely many times.

In summary, we have introduced two forms of texts: text-variables and the text-constant **prim** . Types of texts are again texts. A text declaration is an atomic context declaring a text-variable to be of a specified type.

### 2.1.4   Parametric Equality

In this section we want to formalize a theory of equations over some sort. Even though we only need to work with equations over natural numbers, we want to formalize a version of equality which abstracts from a concrete sort. This makes sense since the notion of equality is of such general quality that it can be used in many different contexts. In the next section we will then illustrate the mechanisms Deva provides to instantiate the parametric equality to the case of natural numbers.

Remember that contexts are the syntactic means of Deva to structure formalizations. Hence we will construct in the following a context to express our formalization of the parametric equality. In order to abstract from a concrete sort, we will formalize equality in terms of a parametric sort $s$. This is expressed in Deva by declaring the text variable $s$ to be of type *sort*. So to begin with we have the following structure:

⟨ A first version of a parametric equality. 2.1.4 ⟩ ≡

**context**  *ParametricEquality* :=

⟦ $s$ : *sort*

; ⟨ Operator declaration.  2.1.4.1 ⟩

; ⟨ Axioms.  2.1.4.2 ⟩

; ⟨ Proof of a simple law.  2.1.4.3 ⟩

⟧

Before continuing, we should make some comments on the abstraction mechanisms of Deva. Principally, Deva provides two abstraction mechanisms: one relating to contexts and the other relating to texts. Context abstraction was alluded to in the previous paragraph. We will discuss it in greater detail in the next section when we demonstrate how the parametric equality is instantiated to the case of natural numbers. Text abstraction is quite similar to $\lambda$-abstraction. However, whereas a $\lambda$-abstraction abstracts a $\lambda$-term over a variable (of some declared typed), Deva generalizes this concept and allows to *abstract a text over a context*. (Remember, that basically a context can be viewed as a structured compilation of declarations and definitions.) The abstraction of a context $c$ over a text $t$ is denoted by $[c \vdash t]$ (read "abstract $c$ over $t$") and is again a text.

The symbol '⊢' is not the same as the "turnstyle" used in logic. It should rather
be seen as a variant of an arrow such as $|\rightarrow$. But it resembles the turnstile in
the sense that inference rules can be formalized as abstractions. Some syntactic
simplifications may be applied to the context $c$, for example enclosing context
brackets may be omitted. The reader has already met an instance of this concept
in Sect. 2.1.2, when we discussed the text

$$\underbrace{\underbrace{[a, b : nat}_{\text{context}} \vdash \underbrace{(a+b)^2 = a^2 + 2 \times a \times b + b^2]}_{\text{text}}}_{\text{text}}$$

as the formalized version of the binomial formula. In this example, we abstract
the text $(a+b)^2 = a^2 + 2 \times a \times b + b^2$ over the context $a, b : nat$ which is a
shorthand notation for $[\![\, a : nat; b : nat \,]\!]$.

Intuitively, text abstraction can be interpreted in a number of different ways.
First of all, a text abstraction can be seen as a functional abstraction of a
function body over a list of parameter declarations [*param decls* ⊢ *fun body*].
Next, a text abstraction can be seen as a scoping construct accompanied by
some declarations local to the scope [*local decls* ⊢ *block body*]. Last, a text
abstraction can also be considered as an inference rule of a conclusion from a list
of hypotheses [*hyps* ⊢ *concls*]; hence, one can read [$c$ ⊢ $t$] also as "from $c$ infer $t$".
In the following we will see examples of each of these different interpretations.

Dual to the notion of abstraction is the notion of *application* without which
abstraction would not be of any use. Deva knows several different ways to denote
an application: Given two texts $t_1$ and $t_2$, the application of $t_1$ to $t_2$ can be written
as $t_1(t_2)$, or $t_1 \,/\, t_2$, or $t_2 \setminus t_1$. To keep the latter two notations apart think of the
application dash to point towards the function $t_1$. Which one of these different
notation one chooses is very much a matter of personal taste and style. Indeed,
the availability of different notation is very helpful for designing developments
which are easy to read. The meaning of an application can be explained in terms
of the interpretation of abstractions. For example, the application of a functional
abstraction to some argument texts yields the function body with the parameters
instantiated to the arguments. Similarly, the application of a scope to some
argument texts yields the body of the scope with the local variables instantiated
to the arguments. Of course, the type conditions expressed in the parameter
declarations or the local declarations have to be obeyed. This type-check is of
particular importance when applying an inference rule to some texts to yield the
conclusion. Each of these texts encodes the proof of the formula corresponding
to its type. The applicability of the inference rule is now guaranteed by checking
that the types of the proof-texts agree with the hypotheses. The precise meaning
of the preceding statements will be defined in the next chapter but we hope that
its intended meaning will become clear while reading this chapter.

**2.1.4.1.** With these comments in mind it should not be difficult to understand
the following formalization of the parametric equality. We interpret equality as
a binary predicate over the parametric sort $s$, i.e., we declare the text-variable

' $=$ ' to be of type $[s; s \vdash prop]$. This is a shorthand notation for $[x, y : s \vdash prop]$, with $x$ and $y$ new. In general, $[x : s \vdash t]$ may be abbreviated by $[s \vdash t]$ provided that $x$ does not occur free in $t$. In order to enhance readability, we declare ' $=$ ' to be an infix operator by pointing out the argument positions.

$\langle$ Operator declaration. 2.1.4.1 $\rangle \equiv$

$(\cdot) = (\cdot) : [s; s \vdash prop]$

This code is used in section 2.1.4.

By this declaration, the application of ' $=$ ' to two arguments of type $s$, say $t$ and $t'$, can be written as $t = t'$ and is of type $prop$.

**2.1.4.2.** The equality is completely characterized by two axioms: reflexivity and substitution. Both of these axioms are formalized as inference rules in the above sense. Reflexivity says that for a given $x$ of type $s$ one may infer $x = x$. Substitution says that one can substitute equal for equal, or more formally, given $x$ and $y$ of type $s$ and a predicate scheme $P$ parameterized over a variable of type $s$ then, provided one has a proof of $x = y$, one may infer $P(y)$ from a proof of $P(x)$. $P(x)$ and $P(y)$ denote the predicates obtained by instantiating the predicate scheme $P$ by $x$ and $y$. This rule is also frequently called "Leibniz rule".

$\langle$ Axioms. 2.1.4.2 $\rangle \equiv$

$[\![ \; refl \quad : [\, x : s \vdash x = x \,]$

$; \; subst \; : [x, y : \; s \,; P : [s \vdash prop\,] \vdash [x = y \vdash \dfrac{P(x)}{P(y)}]]$

$]\!]$

This code is used in section 2.1.4.

Note that to emphasize the inference character of an abstraction, Deva provides for a "two-dimensional" variant of abstraction, i.e., $\dfrac{c}{t}$ is the same as $[c \vdash t]$.

**2.1.4.3.** The reader should carefully study this formalization since it already contains a number of important aspects of Deva formalizations. In order to point out some of the subtleties involved, we will prove within Deva from the given axioms that equality is symmetric. In our framework, this means that we have to exhibit a Deva text whose type represents the symmetry property.

$\langle$ Proof of a simple law. 2.1.4.3 $\rangle \equiv$

$[\![ \; symmetry\_proposition \; := [x, y : \; s \vdash \dfrac{x = y}{y = x}]$

; *symmetry_proof*          := ⟨ Proof of symmetry. 2.1.4.4 ⟩

]

---

This code is used in section 2.1.4.

A *text-definition* is the third atomic context introduced so far. It associates a text with a text-variable and it is most often used to abbreviate lengthy texts. In contrast to a text-declaration which declares a text-variable to be of a certain type, a text-definition just introduces an abbreviation for a text. Consequently, the type of a *defined variable* is the type of the text which the variable is defined to denote.

**2.1.4.4.**    The idea behind the proof basically runs as follows: We want to derive $y = x$ knowing that $x = y$. Consider the predicate $P$ which holds for all those $z$ that are equal to $x$. By reflexivity we know that $P$ holds for $x$ itself and substitution allows then to infer that $P$ holds for $y$, i.e., $y = x$. A little bit closer to Deva we can argue as follows: Given two objects $x$ and $y$ of type $s$. Assume further a proof of $x = y$ or, equivalently, an object of type $x = y$, let's call it *eq*. Define a predicate $P$ over $s$ such that $P$ holds for $z$ if $z = x$. Then, substitution in the case of $x$, $y$, $P$, and *eq* yields a function which maps a proof of $x = x$ to a proof of $y = x$. But $x = x$ is an axiom by reflexivity. Hence we have constructed a proof of $y = x$ starting from a proof of $x = y$. By abstracting over $x$ and $y$ we therefore succeeded in constructing an object, which for any $x$ and $y$ of type $s$ maps a proof of $x = y$ to a proof of $y = x$, or, equivalently, an object of type *symmetry_proposition*. This argument is in an almost one-to-one correspondence to the following Deva text

⟨ Proof of symmetry. 2.1.4.4 ⟩ ≡
$[x, y : s$
$\vdash [ \; eq : \quad x = y$
$\; ; P \; := [ z : s \vdash z = x ]$
$\; \vdash subst \, (x, y, P, eq)$
$\qquad \therefore [ x = x \vdash y = x ]$
$\quad / \; refl(x)$
$\qquad \therefore y = x$
$]$
$\qquad \therefore [ x = y \vdash y = x ]$
$]$
$\quad \therefore \; symmetry\_proposition$

This code is used in section 2.1.4.3.

Several features of the Deva language are used in this proof. First, given two texts $t$ and $t'$ the *judgement* $t \therefore t'$ (read "$t$ of type $t'$") which is again a text carries the same meaning as $t$ but in addition expresses that $t$ is of type $t'$. This is foremost an aid to the reader of Deva formalizations, but also a way to force a type-check by a Deva-checker. Note that application works modulo currying, i.e., the text $[c \vdash [c' \vdash t]]$ and the text $[c; c' \vdash t]$ have the same meaning. The reader should also recognize a number of very important typing principles when studying this example, namely the typing of identifiers, abstractions, and applications: Quite naturally, an identifier has a type as declared. Given an abstraction $[c \vdash t]$ such that $t$ is of type $t'$ then $[c \vdash t]$ is of type $[c \vdash t']$, i.e., an abstraction itself. This typing principle may succinctly be stated as follows:

$$\frac{[c \vdash t \therefore t']}{[c \vdash t] \therefore [c \vdash t']}$$

Given an application $t_1(t_2)$ such that $t_1$ is of type $[a \vdash b]$ and $t_2$ is of type $a$ then $t_1(t_2)$ is of type $b$:

$$\frac{t_1 \therefore [a \vdash b] \quad t_2 \therefore a}{t_1(t_2) \therefore b}$$

However, this principle describes a particular situation only. More generally, $t_1$ may have a dependent type of the form $[x : a \vdash b]$ where $x$ occurs free in $b$. In this case, $t_1(t_2)$ is of type $b[t_2/x]$ where $b[t_2/x]$ denotes $b$ after substituting every free occurrences of $x$ by $t_2$. This typing principle may be stated as follows:

$$\frac{t_1 \therefore [x : a \vdash b] \quad t_2 \therefore a}{t_1(t_2) \therefore b[t_2/x]}$$

It is nothing else than the well-known rule about dependent types

$$\frac{t_1 : (\lambda x : a)b \quad t_2 : a}{t_1(t_2) : b[t_2/x]}$$

We could continue and state (and prove) other properties of equality such as transitivity etc., but instead we want to draw the reader's attention to another important feature of the Deva language which facilitates the use of Deva formalizations to a great extent. Take another look at the proof of the symmetry property given above. There, we had to supply explicitly all parameters $x$, $y$, $P$, and $eq$ to the substitution axiom, even though some of this information is redundant. For example, the names $x$ and $y$ are implicitly contained in the type of $eq$. On the other hand, the definition of the predicate $P$ could be derived from the context and by taking the definition of substitution into account. Having to supply explicitly each parameter in an application leads to tedious repetition of information and forces to construct Deva texts which are hard to read. For this reason, Deva allows to indicate which parameters should be mechanically derivable from the context of an application by distinguishing two ways of declaring a text variable: *explicit declaration* denoted by a colon ':' and *implicit definition* denoted by a question mark '?'. For example, the declarations

$[\![\ refl \quad : [\,x\,?\,s \vdash x = x\,]$

$;\ subst\ : [x, y\,?\ \ s\,;P\,?\,[s \vdash prop\,] \vdash [x = y \vdash \left|\begin{array}{c} P\,(x) \\ \hline P\,(y) \end{array}\right.]]$

$]\!]$

are alternative ways of formalizing the axioms of reflexivity and substitution. The effect on the proof given above is that the applications $subst\,(x, y, P, eq)$ and $refl\,(x)$ can be replaced by $subst\,(eq)$ and $refl$ and that the definition of the predicate $P$ can be omitted. The resulting proof looks as follows

$[x, y :\ s$

$\vdash [\ eq\ : x = y$

$\quad \vdash \quad subst\,(eq)$

$\qquad \therefore [x = x \vdash y = x\,]$

$\qquad\quad /\ refl$

$\qquad \therefore y = x$

$\quad ]$

$\qquad \therefore [x = y \vdash y = x\,]$

$]$

$\quad \therefore\ symmetry\_proposition$

The layout of this proof is slightly different from the one of the previous proof of symmetry: Instead of indenting the judgements we have chosen to out-dent them to give them more prominence. In the rest of this book we will mostly stick to the former style, future versions of DVWEB will allow to freely choose between the two styles.

*Remark.* The claim about mechanical derivation of parameters must in general be taken with great care. In fact, it will turn out later that the pattern-matching process in its full generality amounts to higher-order unification. With respect to the synthesis of parameters, a number of issues have to be faced, such as non-decidability, exponential complexity, and nondeterminism. A usable approach towards pattern-matching in Deva has to impose some restrictions. In the above example, the implicit parameters $x$, $y$, and $P$ can be automatically derived by the Devil system using the type information contained in the judgments. However, deleting (some) of these judgments introduces all the problems of higher-order unification listed above. A formal definition of the implicit notational level of Deva is presented and discussed in the next chapter. The parameter derivation of the Devil system is based on the calculus described in [101].

This concludes the formalization of the parametric equality. The complete formalization which was referred to in Sect. 2.1.2 is given in Fig. 2, where we have added two special instances of the substitution property. In comparison, we show the rules for a parametric equality in a natural deduction setting. As an instructive exercise, the reader is invited to construct proof-texts for the type of *unfold* and *fold*. In Fig. 3, we give a possible solution to the first exercise.

$\langle$ Parametric equality.  2.1.4.5 $\rangle \equiv$

**context** *ParametricEquality* :=

$[\![\ s \qquad\quad : sort$

$;\ (\cdot) = (\cdot) \quad : [s; s \vdash prop\,]$

$;\ refl \qquad\quad : [\,x\,?\,s \vdash x = x\,]$

$;\ symmetry\ : [x, y\,?\,s \vdash \dfrac{\begin{vmatrix} x = y \\ y = x \end{vmatrix}}{}]$

$;\ subst \qquad : [x, y\,?\,s\,;P\,?\,[s \vdash prop\,] \vdash [x = y \vdash \begin{vmatrix} P\,(x) \\ P\,(y) \end{vmatrix}]\,]$

$;\ unfold \qquad : [x, y, z\,?\,s\,;F\,?\,[s \vdash s\,] \vdash [x = y \vdash \begin{vmatrix} z = F(x) \\ z = F(y) \end{vmatrix}]\,]$

$;\ fold \qquad\quad : [x, y, z\,?\,s\,;F\,?\,[s \vdash s\,] \vdash [x = y \vdash \begin{vmatrix} z = F(y) \\ z = F(x) \end{vmatrix}]\,]$

$]\!]$

This code is used in section 2.1.2.

$$\frac{}{x = x}\ refl \qquad\qquad \frac{x = y \quad P(x)}{P(y)}\ subst$$

$$\frac{x = y \quad z = F(x)}{z = F(y)}\ unfold \qquad \frac{x = y \quad z = F(y)}{z = F(x)}\ fold$$

**Fig. 2.** Equality in Deva vs. equality in a natural deduction setting.

## 2.1.5   Natural Numbers

In this section we are concerned with the formalization of a theory of natural numbers as it is needed for the proof of the binomial formula. The formalization is very similar to an equational algebraic specification such as the one shown in Fig. 4. The outline of the Deva formalization is as follows:

$\langle$ Natural numbers.  2.1.5 $\rangle \equiv$

⟨ Proof text for unfold. 2.1.4.6 ⟩ ≡

$[x, y, z ? s$

$; F \quad ? [s \vdash s]$

$\vdash [eq : x = y$

$\vdash [eqq : z = F(x)$

$\vdash subst\ (eq, eqq)$

$\therefore z = F(y)$

$]$

$]$

$]$

**Fig. 3.** Proof text for unfold

**sorts** $nat$

**ops** $\quad 0, 1, 2 :\to nat$

$\qquad succ : nat \to nat$

$\qquad +, \times, \hat{} : (nat, nat) \to nat$

**vars** $\quad m, n : nat$

**eqs** $\quad 1 = succ(0)$

$\qquad 2 = succ(1)$

$\qquad 0 + n = n$

$\qquad succ(n) + m = succ(n + m)$

$\qquad 0 \times n = 0$

$\qquad succ(n) \times m = (n \times m) + m$

$\qquad n \,\hat{}\, 0 = succ(0)$

$\qquad n \,\hat{}\, succ(m) = (n \,\hat{}\, m) \times m$

**Fig. 4.** Abstract data type for natural numbers

**context** $NaturalNumbers :=$

$[\![$ $nat : sort$

$; \langle$ Constructors of natural numbers. 2.1.5.1 $\rangle$

$; \langle$ Equality of natural numbers. 2.1.5.2 $\rangle$

$; \langle$ Operators on natural numbers. 2.1.5.3 $\rangle$

$; \langle$ Properties of the operators on natural numbers. 2.1.5.4 $\rangle$

$]\!]$

This code is used in section 2.1.2.

**2.1.5.1.** First, we declare the two constructor symbols '0' and '*succ*' and define abbreviations for the first two non-zero natural numbers:

⟨ Constructors of natural numbers. 2.1.5.1 ⟩ ≡

$⟦ 0 \quad : \quad nat$

$; \ succ : \ [nat \vdash nat]$

$; 1 \quad := \ succ\,(0)$

$; 2 \quad := \ succ\,(1)$

$⟧$

This code is used in section 2.1.5.

**2.1.5.2.**  We will proceed by defining operations over naturals in an equational style. Hence we have to introduce equality of natural numbers. We will do this by *using* the parametric equality of the previous section and *specializing* it to the case of natural numbers. In general, the *use of a context-variable p* is denoted by **import** $p$. The effect is that a copy of the context $c$ of the definition **context** $p := c$ is included at the point. Thus, we can introduce equality of natural numbers by specializing the context **import** *ParametricEquality* to the text *nat*. In general, the specialization of a context is described by a *context application* $c(t)$ where $c$ is a context and $t$ is a text. The result is a context where the first declared variable is identified with the text $t$, provided that their types agree. Thus, we can introduce equality of natural numbers within the present context by saying

⟨ Equality of natural numbers. 2.1.5.2 ⟩ ≡

**import**  *ParametricEquality* $(nat)$

This code is used in section 2.1.5.

This yields the context *ParametricEquality* in which the text variable $s$ is identified with the text variable *nat*. This example illustrates how a sequential composition $⟦\,x : t; c\,⟧$ can be interpreted as the abstraction of variable $x$ of type $t$ over the context $c$.

**2.1.5.3.**  Now we are able to declare the type of addition, multiplication, and exponentiation, and state their defining properties in an equational style:

⟨ Operators on natural numbers. 2.1.5.3 ⟩ ≡

$⟦ (\cdot) + (\cdot) \quad : [nat; nat \vdash nat]$

$; \ add\_def \ : ⟨\ base \ := [\,n\,?\,nat \vdash 0 + n = n\,]$

$\qquad\qquad , \ recur := [n, m\,?\ nat \vdash succ(n) + m = succ(n + m)]$

$\qquad\qquad ⟩$

$; \ (\cdot) \times (\cdot) \quad : [nat; nat \vdash nat]$

$; \ mult\_def : ⟨\ base \ := [\,n\,?\,nat \vdash 0 \times n = 0\,]$

$\qquad\qquad , \ recur := [n, m\,?\ nat \vdash succ(n) \times m = (n \times m) + m]$

$\qquad\qquad ⟩$

$; (\cdot)^{(\cdot)}$          $: [nat; nat \vdash nat]$
$; exp\_def$     $: \langle base$   $:= [n ? nat \vdash n^0 = 1]$
                $, recur := [n, m ? nat \vdash n^{succ(m)} = n^m \times n]$
                $\rangle$
$]$

This code is used in section 2.1.5.

In the context above we also made use of another construct of Deva: *products*. In general, a product helps to group together related texts, in particular those which denote rules or laws. The different component texts can be (but need not be) named, are separated by comma, and are put between special angle brackets[3]. Each component of a product may be accessed by its name, e.g. *add_def .base*, or by its position, e.g. *add_def .1*. This access operation is called *projection*.

**2.1.5.4.**   In order to prove the binomial formula we will need additional properties of these operations besides the defining properties. These are associativity and commutativity of addition and multiplication.

$\langle$ Properties of the operators on natural numbers. 2.1.5.4 $\rangle \equiv$
$[\![ add\_props$   $: \langle assoc$    $:= [n, m, l ? nat \vdash n + (m + l) = (n + m) + l]$
                $, commut := [n, m ? nat \vdash n + m = m + n]$
                $\rangle$
$; mult\_props : \langle assoc$    $:= [n, m, l ? nat \vdash n \times (m \times l) = (n \times m) \times l]$
                $, commut := [n, m ? nat \vdash n \times m = m \times n]$
                $\rangle$
$; \langle$ Further properties of the operators on natural numbers. 2.1.5.5 $\rangle$
$]\!]$

This code is used in section 2.1.5.

**2.1.5.5.**   And then there is distributivity and two basic properties of doubling and squaring.

$\langle$ Further properties of the operators on natural numbers. 2.1.5.5 $\rangle \equiv$

$distr$    $: \dfrac{n, m, l, k ? nat}{(n + m) \times (l + k) = n \times l + n \times k + m \times l + m \times k}$

$; doubling : [n ? nat \vdash 2 \times n = n + n]$

---

[3] Some readers may have seen publications about Deva using a square bracket notation for products. In fact, this notation has been dropped because it violated the LALR(1) property.

; *squaring* : $[\, n\,?\, nat \,\vdash\, n^2 = n \times n \,]$

See also section 2.1.5.6.

This code is used in section 2.1.5.4.

**2.1.5.6.** Of course, these properties are derived properties in the sense that they can be proven from the axioms defining the natural numbers. Hence we could give proof texts for each one. However, if we want to do this, one axiom is still missing, namely the induction principle for natural numbers. It states that if we are given both, a proof that some predicate $P$ holds for 0, and a proof that one can derive a proof of $P(succ(n))$ assuming a proof of $P(n)$, then we are allowed to conclude that $P$ holds for any natural number. Writing this induction principle as an inference rule, we have

$$(induction) \quad \frac{P(0) \quad \forall n : nat \bullet P(n) \Rightarrow P(succ(n))}{\forall n : nat \bullet P(n)}$$

where $P$ denotes a predicate with a natural number as paramter. Recall that such an axiom is not explicitly given in the specification of an abstract data type. It is usually left implicit and captured by the initial semantics of the abstract data type. Formalizing this axiom in Deva is fairly straightforward. Both implication and universal quantification are captured by means of (text-) abstractions. The implication $P(n) \Rightarrow P(succ(n))$ translates to $[P(n) \vdash P(succ(n))]$ following the interpretation of an abstraction as an inference. The universal quantification is interpretetd as a function which maps a natural number to an proof of $P(n) \Rightarrow P(succ(n))$ (i.e., a text of type $[P(n) \vdash P(succ(n))]$). The complete axiom is thus formalized as follows:

⟨Further properties of the operators on natural numbers. 2.1.5.5⟩+ ≡

; *induction* : $[\, P\,?[nat \vdash prop\,] \vdash \dfrac{P\,(0); [\,n : nat \,\vdash\, [\,P\,(n) \vdash P(succ(n))\,]\,]}{[\,n\,?\,nat \,\vdash\, P(n)\,]}\,]$

Note that this formalization of the induction principle is not the only possible one. Using a theory of predicate calculus, which will become available later, one could write a formalization that is even closer to the usual rule; one would then use quantification over $n$ and implication.

By using this rule we could construct proof texts for the properties declared above, but the proofs are quite technical and involve several lemmas besides the rules themselves. Therefore, we omit these proofs since our main goal is to construct a proof of the binomial formula (see however Sect. 2.2.2 for an inductive proof of the doubling property). Notice that this, i.e. omitting proofs, is both a dangerous as well as a very useful technique of theory formalization. The danger involved in freely declaring rules as axioms and not deriving them from the "real" axioms of the theory is clear, and it is evident that a very disciplined

use of this method is indispensable. On the other hand, the ability to postpone
the proof of rules which are deemed to be true is very useful since otherwise any
formalization tends to get bogged down with tedious and technical proofs which
must be completed before giving a solution to the global problem. Furthermore,
the postponement of proofs allows to mimic a top-down proof style which is
often used in the literature.

This completes the formalization of a theory of natural numbers sufficient for
our purpose. To summarize, this section did not introduce many new concepts of
Deva, but instead applied those introduced in the previous sections to a formalization of a theory of natural numbers which is completely shown in Fig. 5. We
did introduce *context application* and pointed out how it relates to *context abstraction* mentioned in the previous section. Furthermore, we introduced *context
use* and *products*.

### 2.1.6   Proof of the Binomial Formula

This section will finally be concerned with the proof of the binomial formula.
As a first step, we use the theory of natural numbers as it was developed in the
previous section and then give a name to the formula we want to prove:

⟨ Proof of the binomial formula. 2.1.6 ⟩ ≡

**context**  *ProofOfBinomialFormula* :=

⟦ **import**  *NaturalNumbers*

; *thesis* := $[a, b: nat \vdash (a + b)^2 = a^2 + 2 \times (a \times b) + b^2]$

; *proof* := $[a, b: nat \vdash ⟨ \text{Proof body. 2.1.6.1} ⟩] \therefore thesis$

⟧

**2.1.6.1.**   The actual proof now runs exactly like the proof given at the beginning of this chapter. The difference is that the justifications given in braces turn
into operations which transform one step of the derivation into the next. Our
major tools will be the fold and unfold rules of the equality instantiated by the
properties of the arithmetic operations. The shape of *thesis* and the shape of
these two rules motivate the following procedure: we start with the equation
$(a + b)^2 = (a + b)^2$ and transform it to $(a + b)^2 = a^2 + 2 \times (a \times b) + b^2$ by operating on the right hand side of the intermediate equations. So let us begin by
first defining an abbreviation for the (constant) left-hand side of the equations
and by starting the development of the proof with the basic property of squaring

⟨ Proof body. 2.1.6.1 ⟩ ≡

⟦ *LHS* := $(a + b)^2$

⊢ *squaring*

    $\therefore LHS = (a + b) \times (a + b)$

   ⟨ Proof development. 2.1.6.2 ⟩

⟧

**context**  *NaturalNumbers* :=

⟦ *nat*            :    *sort*

; **import**  *ParametricEquality* (*nat*)

; 0              :    *nat*

; *succ*          :    [*nat* ⊢ *nat*]

; 1              :=  *succ* (0)

; 2              :=  *succ* (1)

; (·) + (·)      :    [*nat*; *nat* ⊢ *nat*]

; *add_def*      :    ⦇ *base*  := [ *n* ? *nat* ⊢ 0 + *n* = *n* ]

                    , *recur* := [*n*, *m* ? *nat* ⊢ *succ*(*n*) + *m* = *succ*(*n* + *m*)]

                    ⦈

; (·) × (·)      :    [*nat*; *nat* ⊢ *nat*]

; *mult_def*     :    ⦇ *base*  := [ *n* ? *nat* ⊢ 0 × *n* = 0]

                    , *recur* := [*n*, *m* ? *nat* ⊢ *succ*(*n*) × *m* = (*n* × *m*) + *m*]

                    ⦈

; (·)^{(·)}      :    [*nat*; *nat* ⊢ *nat*]

; *exp_def*      :    ⦇ *base*  := [ *n* ? *nat* ⊢ $n^0 = 1$]

                    , *recur* := [*n*, *m* ? *nat* ⊢ $n^{succ(m)} = n^m \times n$]

                    ⦈

; *add_props*   :    ⦇ *assoc*    := [*n*, *m*, *l* ? *nat* ⊢ *n* + (*m* + *l*) = (*n* + *m*) + *l*]

                    , *commut* := [*n*, *m* ? *nat* ⊢ *n* + *m* = *m* + *n*]

                    ⦈

; *mult_props*  :    ⦇ *assoc*    := [*n*, *m*, *l* ? *nat* ⊢ *n* × (*m* × *l*) = (*n* × *m*) × *l*]

                    , *commut* := [*n*, *m* ? *nat* ⊢ *n* × *m* = *m* × *n*]

                    ⦈

; *distr*        :    $\dfrac{n, m, l, k ?\ nat}{(n + m) \times (l + k) = n \times l + n \times k + m \times l + m \times k}$

; *doubling*     :    [ *n* ? *nat* ⊢ 2 × *n* = *n* + *n*]

; *squaring*     :    [ *n* ? *nat* ⊢ $n^2 = n \times n$]

; *induction*   :    [ *P* ?[*nat* ⊢ *prop*] ⊢ $\dfrac{P\,(0);\, [\, n : nat\ \vdash\, [\, P\,(n) \vdash P(succ(n))\,]\,]}{[\, n\, ?\ nat\ \vdash\ P(n)\,]}$ ]

⟧

**Fig. 5.** Natural Numbers

This code is used in section 2.1.6.

Note that in general a system will only be able to automatically derive the implicitly defined parameters if one supplies sufficient type information. Thus, here *squaring* stands for *squaring* $(n := a + b)$ because we *do* write the result-type $LHS = (a + b) \times (a + b)$. The pattern-matching mechanism synthesizes $n := a + b$ by matching $n^2 = n \times n$ with $(a + b)^2 = (a + b) \times (a + b)$.

**2.1.6.2.**    Next, we apply the distributivity property:

⟨ Proof development. 2.1.6.2 ⟩ ≡
\ *unfold* (*distr*)
    ∴ $LHS = a \times a + a \times b + b \times a + b \times b$

See also sections 2.1.6.3 and 2.1.6.4.

This code is used in section 2.1.6.1.

**2.1.6.3.**    Again, the type information is vital to automatically derive implicit parameters. Next we apply commutativity and associativity rules:

⟨ Proof development. 2.1.6.2 ⟩+ ≡
\ *unfold* (*mult_props.commut*)
    ∴ $LHS = a \times a + a \times b + a \times b + b \times b$
\ *unfold* (*add_props.assoc*)
    ∴ $LHS = a \times a + (a \times b + a \times b) + b \times b$

As an exercise, the reader is encouraged to calculate the concrete instantiations of implicit parameters in the preceding two proof steps.

**2.1.6.4.**    Finally, applying (in reverse) the basic properties of squaring and doubling yields the desired result.

⟨ Proof development. 2.1.6.2 ⟩+ ≡
\ **loop alt** [*fold* (*doubling*), *fold* (*squaring*)]
    ∴ $LHS = a^2 + 2 \times (a \times b) + b^2$

Here, two new constructs of Deva are introduced. If $t_1, \ldots, t_n$ are inference rules, then **alt** $[t_1, \ldots, t_n]$ is a text which, when applied to another text $t$, chooses the rule to apply depending on the type of $t$. In this sense **alt** $[t_1, \ldots, t_n]$ is another inference rule. Given an inference rule $t$, the intended meaning of the text **loop** $t$ is to iterate the application of $t$. The number of iterations is determined by a proof search process which, in this case, is guided by the final judgment, i.e., the goal. These two constructs are very useful for building so-called *tactics*. In the following discussions, our use of the term "tactic" will always refer to texts based on **alt** and **loop**.

*Remark.* It is clear that constructs such as **alt** and **loop** cannot be used arbitrarily, if combinatorial explosion is to be avoided. The above kind of use, however, is quite reasonable since the iteration is unfolded into application sequences with three steps only, one possible sequence is shown below. Moreover, such a use of tactics is quite frequent in the conceptual development of a proof by a human.

$\backslash fold(doubling)$

$\quad \therefore LHS = a \times a + 2 \times (a \times b) + b \times b$

$\backslash fold(squaring)$

$\quad \therefore LHS = a^2 + 2 \times (a \times b) + b \times b$

$\backslash fold(squaring)$

$\quad \therefore LHS = a^2 + 2 \times (a \times b) + b^2$

This concludes the proof of the binomial formula formalized with Deva and our first introduction to the language. The proof is completely shown in Fig. 6 where we have once more chosen to outdent the judgements in order. Comparing this formalized version of the proof to the rigorous one given earlier (reiterated for convenience), we can make several observations. The assumptions left implicit in the equational proof have been made explicit by basing the proof on an equational theory of natural numbers. Next, the transformation rules used in the proofs are explicitly stated so that the proof can be checked by a machine. Finally, whereas in the rigorous proof the terms themselves are transformed, in the formalized proof equations are transformed. This is precisely how equational reasoning has been formalized, i.e., given equations are transformed into new equations. This formalization style was chosen because of its clarity and logical adequacy. However it does not directly correspond to equational reasoning as used in the informal proof[4]. But still, the overall structure of the equational proof has been maintained and the notational overhead — which is the price to be paid for complete formality — has been kept at a low level thanks to the different constructs of the Deva language.

To be fair, one must also acknowledge the effort of formalizing the theories of parametric equality and natural numbers. Here we argue that such theories should be available in a library of formalized mathematics. Chapter 4 presents some theories such a library might contain.

## 2.2  Elements of Proof Design

The first part of this chapter has been mainly concerned with a formal proof of a particular kind, a proof by equational calculation. In this second part we will

---

[4] One could formalize equality in a different manner which would then lead to an even closer match with the rigorous proof. The basic idea is to interpret $[a \vdash b]$ as $a = b$, assuming $a$ and $b$ are not propositions, and then to build a set of rules (such as substitution) around this formalization of equality. However, this would raise several problems which we do not want to discuss at this time.

---

**context** *ProofOfBinomialFormula* :=

[ **import** *NaturalNumbers*

; *thesis* := $[a, b : nat \vdash (a+b)^2 = a^2 + 2 \times (a \times b) + b^2]$

; *proof* := $[a, b : nat$

$\vdash [ LHS := (a+b)^2$

$\vdash$    *squaring*

$\therefore LHS = (a+b) \times (a+b)$

$\backslash$ *unfold(distr)*

$\therefore LHS = a \times a + a \times b + b \times a + b \times b$

$\backslash$ *unfold(mult_props.commut)*

$\therefore LHS = a \times a + a \times b + a \times b + b \times b$

$\backslash$ *unfold(add_props.assoc)*

$\therefore LHS = a \times a + (a \times b + a \times b) + b \times b$

$\backslash$ **loop alt** $[fold(doubling), fold(squaring)]$

$\therefore LHS = a^2 + 2 \times (a \times b) + b^2$

]

]

$\therefore$ *thesis*

]

---

$(a+b)^2$

$=$    { definition of squaring }

$(a+b) \times (a+b)$

$=$    { distributivity }

$a \times a + a \times b + b \times a + b \times b$

$=$    { commutativity }

$a \times a + a \times b + a \times b + b \times b$

$=$    { definition of squaring, definition of "doubling" }

$a^2 + 2 \times a \times b + b^2$

**Fig. 6.** Deva Proof of the Binomial Formula vs. Equational Proof

give an introduction to how to describe proofs in a variety of other styles in the Deva language. Formal development methods usually do not enforce particular proof styles, but it is well known that the choice or the combination of styles to structure and describe proofs may greatly influence the elegance, readability, and reusability of proofs. This discussion about structured proving is quite similar to the discussion about structured programming. The proof styles explained in this introduction are used later in the case studies and give a good idea of how to structure formal proofs in Deva.


### 2.2.1   Transitive Calculations


In general, a calculational proof style is characterized by the transformation of
expressions using algebraic identities. Such identities need not be restricted to
equations but can involve any transitive binary relation such as preorders. The
usefulness and power of such generalized *calculations* has been aptly stated by
Dijkstra and Feijen in [34] where they also introduce the proof layout (due to
Feijen) which we have used to present the introductory example (cf. also [35]).
A context formalizing the axioms of a preorder in Deva is shown below. Preorders often serve to perform approximations in proofs. Well-known examples
are the less-than relation over the naturals or the set-inclusion relation. When
viewed sufficiently abstract, data-refinement can also be viewed as a preorder
(cf. Chap. 5.3).


**context**  *Preorder* :=

$[\![\ s \qquad\qquad : sort$

$;\ (\cdot) \le (\cdot)\ \ :[\,s;s \vdash prop\,]$

$;\ pre\_refl\ \ \ :[\,x\,?\,s \vdash x \le x\,]$

$;\ pre\_trans\ :[\,x,y,z\,?\ s \vdash \dfrac{x \le y; y \le z}{x \le z}\,]$

$]\!]$


### 2.2.2   Lemmas and Tactics


A very common way to structure a proof is to identify a number of lemmas that
can be proven in separation and composed within the "main part" of the proof.
As an example, we show a proof of the doubling property of natural numbers
(cf. Sect. 2.1.5) which is structured into two lemmas and one main part.


$\langle$ Doubling proof. 2.2.2 $\rangle \equiv$

**context**  *DoublingProof* :=

$[\![$ **import**  *NaturalNumbers*

$;\ transform \qquad := \langle\,Transformation\ tactics.\ 2.2.2.1\,\rangle$

$;\ doubling\_thesis := [\,n\,?\,nat \vdash n \times 2 = n + n\,]$

; $succ\_lemma$    $:= [\, n \; ? \; nat$
$\qquad\qquad\quad \vdash refl$
$\qquad\qquad\qquad \therefore succ(n) = succ(n)$
$\qquad\qquad\qquad \backslash \mathbf{loop} \; transform \, . \, add \, . \, def$
$\qquad\qquad\qquad \therefore succ(n) = 1 + n$
$\qquad\qquad\qquad \backslash transform. \, add \, . \, ac$
$\qquad\qquad\qquad \therefore succ(n) = n + 1$
$\qquad\qquad\quad ]$

; $base\_lemma$    $:= mult\_def \, . \, base$
$\qquad\qquad\qquad \therefore 0 \times 2 = 0$
$\qquad\qquad\qquad \backslash transform. \, add \, . \, def$
$\qquad\qquad\qquad \therefore 0 \times 2 = 0 + 0$
$\qquad\qquad\qquad \therefore doubling\_thesis(n := 0)$

; $recur\_lemma$    $:= [\, m \quad : \quad nat$
$\qquad\qquad\quad ; hyp \quad : \quad m \times 2 = m + m$
$\qquad\qquad\quad ; LHS \; := \; succ\,(m) \times 2$
$\qquad\qquad\quad \vdash refl$
$\qquad\qquad\qquad \therefore LHS = succ(m) \times 2$
$\qquad\qquad\qquad \backslash unfold(mult\_def\,.recur)$
$\qquad\qquad\qquad \therefore LHS = m \times 2 + 2$
$\qquad\qquad\qquad \backslash unfold(hyp)$
$\qquad\qquad\qquad \therefore LHS = (m + m) + 2$
$\qquad\qquad\qquad \backslash unfold(succ\_lemma)$
$\qquad\qquad\qquad \therefore LHS = (m + m) + (1 + 1)$
$\qquad\qquad\qquad \backslash \mathbf{loop} \; transform \, . \, add \, . \, ac$
$\qquad\qquad\qquad \therefore LHS = (m + 1) + (m + 1)$
$\qquad\qquad\qquad \backslash \mathbf{loop} \; fold \, (succ\_lemma)$
$\qquad\qquad\qquad \therefore LHS = succ(m) + succ(m)$
$\qquad\qquad\quad ]$

$$\therefore [\, m : nat \; \vdash \frac{m \times 2 = m + m}{succ\,(m) \times 2 = succ(m) + succ(m)}\,]$$

$$\therefore [\, m : nat \; \vdash \frac{doubling\_thesis\,(n := m)}{doubling\_thesis\,(n := succ(m))}\,]$$

; $proof$          $:= induction\,(base\_lemma, recur\_lemma)$
$\qquad\qquad\qquad \therefore doubling\_thesis$

$]$

The proof strategy is to proceed by induction: The induction base and the induction step are stated separately in a calculational style and are then used as arguments to the induction rule over naturals. Note the local assumption of the induction hypothesis in the induction step. Note also the iterated application of two simple tactics, in the form of alternatively applicable rules, in the second lemma (*recur_lemma*). Tactical aspects have already been discussed in the proof of the binomial formula. The tactic *transform . add .def* transforms by using the defining properties of addition whereas the tactic *transform . add .ac* transforms by using associativity or commutativity of addition. Tactics like these can be defined by using alternatives to express a choice and products to group rules. Displayed as an excerpt of a collection of tactic definitions, the definition of the two tactics explained above looks as follows:

⟨ Transformation tactics. 2.2.2.1 ⟩ ≡
◁ *add* := ◁ *def* := **alt** [*unfold*(*add_def.base*) , *fold*(*add_def.base*)
                        , *unfold*(*add_def.recur*), *fold*(*add_def.recur*)
                        ]

      , *ac* := **alt** [*unfold*(*add_props.commut*), *fold*(*add_props.assoc*)
                        , *unfold*(*add_props.assoc*)
                        ]

      ▷

▷

This code is used in section 2.2.2.

Similar to tactics, lemmas may quite naturally be structured over several levels. For example the doubling property can be seen as a lemma in the proof of the binomial formula.

Our experimentation with completely formal developments quickly demonstrated to us the enormous impact of the design of definitions and laws and their structuring into theories and tactics on the complexity of developments. This is not at all a new insight but a well-known phenomenon of any mathematical discipline, i.e., definitions and theories *structure* the mathematical thinking. There are no formal rules that define what is a "good" definition or a "good" structuring. In a completely formalized framework as ours, a reasonable guideline is to ensure within developments a high degree of reuse of established properties from the particular application theory. In addition, tactics may allow to summarize often repeated patterns of subsequent proof steps into a single step. As a benefit, using well-structured theories and tactics, one can often avoid to "start from scratch" in proofs. The situation is comparable to that of an applied mathematician who often works by applying specialized rulesets for the economic solution of particular classes of problems. The case studies in this book give an, admittedly limited, illustration of these issues.

### 2.2.3   Local Scope

Local scope in proofs arises when one introduces the premises of a theorem as local hypotheses, from which the conclusion of the theorem is to be derived. We will illustrate local scope in proofs in the context of a very simple logic. The following context shows a formalization of *Minimal Logic*, a propositional logic about implications.

⟨ Minimal Logic. 2.2.3 ⟩ ≡
**context**  *MinimalLogic* :=
[ (·) ⇒ (·) : [*prop*; *prop* ⊢ *prop*]

; *intro*      : [*a*, *b* ? *prop* ⊢ $\dfrac{\left| [a \vdash b] \right.}{a \Rightarrow b}$]

; *elim*      : [*a*, *b* ? *prop* ⊢ $\dfrac{a \Rightarrow b}{\left| [a \vdash b] \right.}$]

; ⟨ A proof with local scope. 2.2.3.1 ⟩
]

In Minimal Logic, the only connector for propositions is the implication symbol ' ⇒'. Two inference rules allow the introduction and elimination of implications. An implication $a \Rightarrow b$ can be introduced if it was possible to prove $b$ assuming $a$, i.e., to prove $[a \vdash b]$. The elimination rule is very natural and widely known under the name *modus ponens*.

**2.2.3.1.**   Given the formalization of Minimal Logic, we will now prove that ⇒ distributes from the left over itself, i.e.,

⟨ A proof with local scope. 2.2.3.1 ⟩ ≡
    *thesis* := [*a*, *b*, *c* : *prop* ⊢ $(a \Rightarrow (b \Rightarrow c)) \Rightarrow ((a \Rightarrow b) \Rightarrow (a \Rightarrow c))$]
 ; *proof* := [*a*, *b*, *c* : *prop*
                       ⊢⟨ First scope. 2.2.3.2 ⟩
                       ]
                 ∴ *thesis*

This code is used in section 2.2.3.

**2.2.3.2.**   As explained above, we have to construct an abstraction with type $[a \Rightarrow (b \Rightarrow c) \vdash (a \Rightarrow b) \Rightarrow (a \Rightarrow c)]$. Application of the introduction rule yields the desired thesis. The proof of $(a \Rightarrow b) \Rightarrow (a \Rightarrow c)$ is deferred to a second scope.

⟨ First scope. 2.2.3.2 ⟩ ≡
[ $h_1$ : $a \Rightarrow (b \Rightarrow c)$
⊢⟨ Second scope. 2.2.3.3 ⟩
]

$$\therefore [a \Rightarrow (b \Rightarrow c) \vdash (a \Rightarrow b) \Rightarrow (a \Rightarrow c)]$$

$\backslash$ *intro*

$$\therefore (a \Rightarrow (b \Rightarrow c)) \Rightarrow ((a \Rightarrow b) \Rightarrow (a \Rightarrow c))$$

This code is used in section 2.2.3.1.

**2.2.3.3.** As it happens, in this second scope we have to prove another implication, namely $(a \Rightarrow b) \Rightarrow (a \Rightarrow c)$. Again, we do this by constructing an abstraction, this time with type $[a \Rightarrow b \vdash a \Rightarrow c]$ and apply the introduction rule.

$\langle$ Second scope. 2.2.3.3 $\rangle \equiv$

$[\, h_2 : a \Rightarrow b$

$\vdash \langle$ Third scope. 2.2.3.4 $\rangle$

$]$

$$\therefore [a \Rightarrow b \vdash a \Rightarrow c]$$

$\backslash$ *intro*

$$\therefore (a \Rightarrow b) \Rightarrow (a \Rightarrow c)$$

This code is used in section 2.2.3.2.

**2.2.3.4.** Well, the second scope called for the proof of another implication, $a \Rightarrow c$.

$\langle$ Third scope. 2.2.3.4 $\rangle \equiv$

$[\, h_3 : a$

$\vdash \langle$ Core of the proof. 2.2.3.5 $\rangle$

$]$

$$\therefore [a \vdash c]$$

$\backslash$ *intro*

$$\therefore a \Rightarrow c$$

This code is used in section 2.2.3.3.

**2.2.3.5.** Here we are at the core of the proof: we have to construct a proof of $c$. At this point we have introduced three assumptions $h_1$, $h_2$, and $h_3$; they must suffice to produce a proof of $c$. This proof turns out to be quite simple to construct: The elimination rule can be applied twice to the assumptions giving *elim* $(h\_2, h\_3) \therefore b$ and *elim* $(h_1, h_3) \therefore b \Rightarrow c$. This suggests a third application to combine these results into a proof of $c$.

$\langle$ Core of the proof. 2.2.3.5 $\rangle \equiv$

$$elim\,(elim(h_1, h_3), elim(h_2, h_3)) \mathrel{\therefore} c$$

This code is used in section 2.2.3.4.

The proof as a whole is shown in Fig. 7. At this point, it is again worthwhile to compare the Deva formalization of Minimal Logic to a more conventional one, e.g. in a sequent calculus style [43]. Such a formalization is given in Fig. 8. Note that, except for structural differences, the proof proceeds in much the same way as the proof in Deva: first $c$ is constructed by using the assumptions and modus ponens (i.e., the '*elim*' rule) and then the assumptions are discharged by successive applications of the '*intro*' rule. The schema variables $A$ and $B$ are represented by the (implicit) parameters of the rules. The structural differences are due to the different handling of assumptions. In the sequent calculus, the assumptions are recorded in a list which is part of a sequent. In the Deva formalization, assumptions are introduced by a (text-) abstraction; in its scope they may be used; they are discharged by a (text-) application. Thus, Deva has already key concepts of a deduction calculus "built-in". A certain overhead with respect to the sequent-based proof results from the fact that Deva enforces the use of explicit pointers ($h_1$, $h_2$, $h_3$) to assumptions, such pointers are not needed in Fig. 8.

## 2.2.4  Composition of Inference Rules

Most of the proofs illustrated so far have been of a general style characterized by successive applications and judgments:

$$
\begin{aligned}
&initialize\\
&\qquad \mathrel{\therefore} result_1\\
&\backslash\; law_1\\
&\qquad \mathrel{\therefore} result_2\\
&\backslash\; law_2\\
&\qquad \mathrel{\therefore} result_3\\
&\qquad \cdots
\end{aligned}
$$

The text *initialize* describes an application of a given axiom used to initialize the reasoning chain. The texts $law_1$, $law_2$, ..., describe (directly given or derived) inference laws. The typing condition requires that *initialize* has type $result_1$, $law_1$ has type $[result_1 \vdash result_2]$, and $law_2$ has type $[result_2 \vdash result_3]$.

Another possible style to describe proofs is characterized by pure *composition* of inference rules, rather than application of inference rules to an initial axiom. The general shape of a compositional proof looks as follows:

$$
\begin{aligned}
&law_0\\
&\qquad \mathrel{\therefore} [result_0 \vdash result_1]\\
&\diamondsuit\; law_1\\
&\qquad \mathrel{\therefore} [result_0 \vdash result_2]\\
&\diamondsuit\; law_2\\
&\qquad \mathrel{\therefore} [result_0 \vdash result_3]\\
&\qquad \cdots
\end{aligned}
$$

$$
\begin{array}{l}
[a, b, c : prop \\
\quad \vdash [\, h_1 \ : a \Rightarrow (b \Rightarrow c) \\
\qquad \vdash [\, h_2 \ : a \Rightarrow b \\
\qquad\quad \vdash [\, h_3 \ : a \\
\qquad\qquad \vdash elim\,(elim(h_1, h_3), elim(h_2, h_3)) \therefore c \\
\qquad\quad ] \\
\qquad\qquad\quad \therefore [a \vdash c] \\
\qquad\quad \backslash\ intro \\
\qquad\qquad\quad \therefore a \Rightarrow c \\
\qquad ] \\
\qquad\quad \therefore [a \Rightarrow b \vdash a \Rightarrow c] \\
\qquad \backslash\ intro \\
\qquad\quad \therefore (a \Rightarrow b) \Rightarrow (a \Rightarrow c) \\
\quad ] \\
\qquad \therefore [a \Rightarrow (b \Rightarrow c) \vdash (a \Rightarrow b) \Rightarrow (a \Rightarrow c)] \\
\quad \backslash\ intro \\
\qquad \therefore (a \Rightarrow (b \Rightarrow c)) \Rightarrow ((a \Rightarrow b) \Rightarrow (a \Rightarrow c)) \\
] \\
\quad \therefore thesis
\end{array}
$$

**Fig. 7.** A proof in Minimal Logic

As a difference to the application-oriented style, the intermediate types are inferences between results, rather than just results. In addition, a new operation is introduced, the *cut* ('$\diamond$') of two texts. The purpose of the cut is to compose two texts in a functional way. This is reflected by the following typing rule for the cut:

$$
\frac{t_1 \therefore [a \vdash b] \quad t_2 \therefore [b \vdash c]}{t_1 \diamond t_2 \therefore [a \vdash c]}
$$

However, this rule describes a particular situation only. More generally, $t_2$ can have a type of the form $[x : b \vdash c]$ in which $x$ occurs free in $c$. The typing rule is generalized as follows:

$$
\frac{t_1 \therefore [a \vdash b] \quad t_2 \therefore [x : b \vdash c]}{t_1 \diamond t_2 \therefore [a \vdash c[t_1/x]]}
$$

Note the similarity to the typing rule for applications (Sect. 2.1.4). One can view the application-oriented style of development as a special case of compositional development by assuming that $result_0$ corresponds to *true*, i.e., a formula that always holds. In this view *initialize* is associated with $law_0$.

$$\frac{}{\Gamma \cup \{A\} \vdash A} \; \text{axiom} \qquad \frac{\Gamma \cup \{A\} \vdash B}{\Gamma \vdash A \Rightarrow B} \; \text{intro} \qquad \frac{\Gamma \vdash A \Rightarrow B \quad \Delta \vdash A}{\Gamma \cup \Delta \vdash B} \; \text{elim}$$

$$\frac{\dfrac{}{a \Rightarrow (b \Rightarrow c) \vdash a \Rightarrow (b \Rightarrow c)} \; \text{ax} \quad \dfrac{}{a \vdash a} \; \text{ax}}{a \Rightarrow (b \Rightarrow c), a \vdash b \Rightarrow c} \; \text{elim} \quad \dfrac{\dfrac{}{a \Rightarrow b \vdash a \Rightarrow b} \; \text{ax} \quad \dfrac{}{a \vdash a} \; \text{ax}}{a \Rightarrow b, a \vdash b} \; \text{elim}$$

$$\frac{a \Rightarrow (b \Rightarrow c), a \Rightarrow b, a \vdash c}{\dfrac{a \Rightarrow (b \Rightarrow c), a \Rightarrow b \vdash a \Rightarrow c}{\dfrac{a \Rightarrow (b \Rightarrow c) \vdash (a \Rightarrow b) \Rightarrow (a \Rightarrow c)}{\vdash (a \Rightarrow (b \Rightarrow c)) \Rightarrow ((a \Rightarrow b) \Rightarrow (a \Rightarrow c))} \; \text{intro}} \; \text{intro}} \; \text{intro}$$

**Fig. 8.** Minimal Logic in a sequent calculus

**2.2.4.1.** As a simple illustration of the cut consider again the substitution law for equality (Sect. 2.1.4). For some reason, one may wish to derive a "reversed" form of substitution in which $x$ and $y$ are exchanged in the conclusion:

$\langle$ Reverse Substitution. 2.2.4.1 $\rangle \equiv$

$rsubst := \langle$ Derivation of reverse substitution. 2.2.4.2 $\rangle$

$$\therefore [x, y \,?\; s \,; P \,? \, [s \vdash prop] \vdash [x = y \vdash \left| \frac{P(x)}{P(y)} \right| ]]$$

**2.2.4.2.** Reversed substitution can be derived quite simply by composing symmetry with substitution

$\langle$ Derivation of reverse substitution. 2.2.4.2 $\rangle \equiv$

$symmetry \diamond subst$

This code is used in section 2.2.4.1.

This proof is rather intuitive to read: reversed substitution with an equation amounts to a reverse of the equation (*symmetry*) followed by (un-reversed) substitution. When studying this example, the reader will probably wonder how it technically works, given the presence of all the implicit definitions in the types of *symmetry* and *subst*. In fact, the proof has used *implicit notation* again: One design goal of Deva was to allow to reduce the formal noise in a proof as much as possible.

**2.2.4.3.** A fully explicit version of the proof of reversed substitution extracts and properly coordinates the implicit parameters of symmetry and substitution:

$\langle$ Derivation of reverse substitution$'$. 2.2.4.3 $\rangle \equiv$

$[x, y ? s$

$; P ? [s \vdash prop]$

$\vdash symmetry (x := x, y := y) \diamond subst(x := y, y := x, P := P)$

$]$

This example illustrates how implicit notation may not only omit arguments but also parameterizations.

Note that the explicit form of the proof of reversed substitution has introduced a variant of the application notation. In a *named application* $t_1(x := t_2)$ the formal parameter $x$ in $t_1$ is instantiated to $t_2$. In particular $x$ need *not* be the first formal parameter of $t_1$. In the above illustration we could have equally well written *symmetry* $(y := y, x := x)$. The typing properties of named application are obvious variants of the properties of application and therefore omitted at this point. The infix notations for application are extended to named application, i.e., $t_2 \setminus t_1$ **at** $x$ and $t_1$ **at** $x \, / \, t_2$ stand for $t_1(x := t_2)$.

**2.2.4.4.** It is also possible to derive reversed substitution without using the cut. Instead, one could make an additional assumption and use a sequence of applications:

⟨Derivation of reverse substitution''. 2.2.4.4⟩ ≡

$[x, y ? s$

$; P ? [s \vdash prop]$

$\vdash[ eq : x = y$

$\vdash eq$

$\therefore x = y$

$\setminus symmetry$

$\therefore y = x$

$\setminus subst$

$\therefore [P(y) \vdash P(x)]$

$]$

$]$

This proof is easier to understand but less economical than the compositional one (even with all judgments erased). In general, the choice which style to use seems to depend on the familiarity with the Deva machinery. The experienced user is likely to recognize situations which call for compositional proofs.

## 2.2.5 Backward Direction

An interesting feature of the compositional style is the fact that it becomes possible to describe proofs in a *backward direction*, i.e., not starting from axioms but from the thesis to be proven. The shape of a simple backward proof can be presented in a compositional style:

$$
\begin{aligned}
&[x : result_3 \vdash x] \\
&\qquad \therefore [result_3 \vdash result_3] \\
&\Diamond\ law_2 \\
&\qquad \therefore [result_2 \vdash result_3] \\
&\Diamond\ law_1 \\
&\qquad \therefore [result_1 \vdash result_3] \\
&\Diamond\ law_0 \\
&\qquad \therefore [result_0 \vdash result_3]
\end{aligned}
$$

The starting point is the trivial proof text $[x : result_3 \vdash x]$ of the fact that the intended result $result_3$, i.e., the goal of the backward proof, is implied by itself. Using the cut in reversed notation, i.e., $t_2 \Diamond t_1$ standing for $t_1 \Diamond t_2$, the goal $result_3$ is then gradually reduced to the new goal $result_0$. Indeed, if $result_0$ is a proposition which is evidently true, then this procedure has produced a proof of the goal.

Backward proofs are usually more complex than the above, idealized, shape. In particular there may not only be one but several subgoals in the intermediate stages of the proof. This situation can be handled by using the *named cut*, a variant of the cut analogous to the named application. In a named cut $t_1 \Diamond t_2$ at $x$, the text $t_1$ is cut with $t_2$ at the declaration of $x$. For example, imagine an intermediate situation of a backward proof with two subgoals $a$ and $b$, in which the subgoal $b$ is to be reduced to a new subgoal $b'$:

$$
\frac{t_1 \therefore [y' : b' \vdash b] \quad t_2 \therefore [x : a; y : b \vdash c]}{t_1 \Diamond t_2 \text{ at } y \therefore [x : a; y' : b' \vdash c]}
$$

To avoid problems with bound variables, the named cut requires that $x$ and $y'$ do not occur free in $b$. Naturally, the named cut also exists in a reversed version, $t_2$ at $y \Diamond t_1$.

## 2.3   Further Constructs

In order to give a complete overview over Deva, in this last part of the introduction we will point out all those aspects of Deva which have not yet been mentioned.

First of all, the argument position in an application, a cut, or a projection can not only be specified by its name but also by a number. For example, $t_1(2 := t_2)$ describes the instantiation of the *second* declaration (or implicit definition) of $t_1$ by $t_2$.

A product may be created by replacing a particular component in a copy of a given product. This *product modification* is written $t_1$ **where** $x := t_2$ and is characterized by the following typing property:

$$
\frac{t_1 \therefore \langle x := a, y := b \rangle \quad t_2 \therefore b'}{t_1 \text{ where } y := t_2 \therefore \langle x := a, y := b' \rangle}
$$

The position to be modified may also be specified by a number.

Two constructs, *sums* and *case distinctions*, are fundamental to the alternative construct which we already used several times. Text¿Case distinction For instance, assuming $f : [a \vdash b]$ and $g : [c \vdash d]$, the alternative **alt** $[f, g]$ is actually implicit notation for the explicit expression:

$$
\begin{array}{l}
[\; x : \langle a \mid c \rangle \\
\;\vdash \textbf{case } x \textbf{ of } [f, g] \\
\;]
\end{array}
$$

The type $\langle a \mid c \rangle$ is a binary sum describing the argument types that the alternative accepts. But the Deva constructs we know at this point do not allow to name any text which has this type. Some sort of "tagging-device" is needed in order to inject a text into a sum. In Deva it is known as the *ghost operator* '^'. Assuming $t$ of type $c$, the text $\langle \hat{\ } a \mid t \rangle$, for example, has $\langle a \mid c \rangle$ as its type. In other words, the special sum $\langle \hat{\ } a \mid t \rangle$ injects $t$ into the sum of texts of type $\langle a \mid c \rangle$. In general, this special form of sums are called *injections*. A case distinction **case** $t_1$ **of** $t_2$ allows to select from a number of alternatives functions in $t_2$, depending on the kind of injection applied to the argument $t_1$.

Finally, the situation often arises that one defines a context and immediately afterwards imports it. For this, Deva provides the shorthand notation '**import context** $c := \ldots$' which is the same as '**context** $c := \ldots$ ; **import** $c$'.

In this chapter, we hope to have demonstrated that the expression in Deva of classical formalizations (equational reasoning, abstract data types, induction, natural deduction, sequent deductions, etc.) is systematic, simple, and readable, and does not unreasonably "blow up".

# 3    Stepwise Definition of Deva

This chapter presents a complete formal definition of the Deva language. A substantial effort has been made to structure this quite technical presentation and the reader may consult the examples of the previous chapter to obtain aryintuition about the constructs.

As already hinted in the introduction, Deva consists of a fully explicit level of notation, the *explicit part*, and a partially implicit level of notation, the *implicit part*. First, two small examples are presented and explained by appealing to the reader's intuition. Throughout the chapter, the examples will be explained in depth as an illustration of the formal definition of Deva. The chapter will then proceed by defining a kernel of the explicit part. This kernel is a quite simple sublanguage of Deva which nevertheless already introduces all essential semantic relations of Deva. Then this kernel is systematically extended by the definition of the remaining Deva operations. Based on the definition of the explicit part, the third section defines the implicit part of Deva. Finally, some mathematical properties of these definitions are presented and illustrated.

## 3.1    Two Examples

The first example (Fig. 9) presents a proof of Syllogism (i.e., transitivity of implication) within a formalization of Minimal Logic which slightly differs from the one presented in Sect. 2.2.3. The formalization of minimal logic is imported and Syllogism is proven. The proof runs as follows: assume, for arbitrary propositions $a$, $b$, and $c$, that $a \Rightarrow b$ and $b \Rightarrow c$ holds. Using the elimination rule on each of these assumptions, one infers that $b$ holds under the assumption of $a$ and $c$ holds under the assumption of $b$. By resolving the intermediate formula $b$, one infers that $c$ holds under the assumption of $a$. The proof of $a \Rightarrow c$ is then obtained by applying the introduction rule.

As a second example, an almost trivial simplification tactic for boolean expressions is described and illustrated in Fig. 10. Two simplification rules are considered:

$$(a \wedge true) \text{ is replaced by } a \quad \text{and} \quad (a \vee false) \text{ is replaced by } a.$$

A simplification tactic (*simplify*) is defined which may alternatively choose between one of these rules. Its use is illustrated on a simple example in which a proof $u$ of $(b \wedge true) \vee false$ is transformed into a proof of $b$.

## 3.2    The Explicit Part: Kernel

Quite obviously, some parts of the examples of Sect. 3.1 were not presented in a completely explicit way. For example, in the Syllogism proof some parameters of the introduction and elimination rules were not instantiated. Similarly, the second example provided only rather rough information about the simplification itself. Both examples have been expressed using the implicit part of Deva. But

**context** *Syllogism* :=
[ **context** *MinimalLogic* :=
  [ *prop*      : **prim**
  ; $(\cdot) \Rightarrow (\cdot)$ : [*prop*; *prop* ⊢ *prop*]
  ; *rules*     : ⟨ *intro* := [$p, q$ ? *prop* ⊢ [[$p$ ⊢ $q$] ⊢ $p \Rightarrow q$]]
                  , *elim* := [$p, q$ ? *prop* ⊢ [$p \Rightarrow q$ ⊢ [$p$ ⊢ $q$]]]
                  ⟩
  ]
; **import**  *MinimalLogic*
; *thesis* := [$a, b, c$ ? *prop* ⊢ [$a \Rightarrow b$; $b \Rightarrow c$ ⊢ $a \Rightarrow c$]]
; *proof*  := [$a, b, c$ ? *prop* ; $asm_1$ : $a \Rightarrow b$; $asm_2$ : $b \Rightarrow c$
            ⊢ (*rules. elim* ($asm_1$) ∴ [$a$ ⊢ $b$])
              ⋄ (*rules. elim* ($asm_2$) ∴ [$b$ ⊢ $c$])
                ∴ [$a$ ⊢ $c$]
              \ *rules. intro*
                ∴ $a \Rightarrow c$
            ]
              ∴ *thesis*
]

**Fig. 9.** Proof of Syllogism in Minimal Logic

**context** *Simplification* :=
[ *bool*             : **prim**
; *true, false*      : *bool*
; $(\cdot) \wedge (\cdot), (\cdot) \vee (\cdot)$ : [*bool*; *bool* ⊢ *bool*]
; *laws*             : ⟨ *and* := [$a$ ? *bool* ⊢ [$a \wedge true$ ⊢ $a$]]
                       , *or*  := [$a$ ? *bool* ⊢ [$a \vee false$ ⊢ $a$]]
                       ⟩
; *simplify*         := [$a$ ? *bool* ⊢ **alt** [*laws. and* ($a := a$)]
                                       , *laws. or* ($a := a$)
                                     ]
; *simplification*   := [ $b$ : *bool*
                       ; $u$ : ($b \wedge true$) $\vee$ *false*
                       ⊢ **loop** *simplify*($u$) ∴ $b$
                       ]
]

**Fig. 10.** Simplification of boolean expressions

before we can explain and understand the implicit part, we will turn our attention to the explicit part. In Sect. 3.5 it will be shown how the explicit part serves as a means to "explain" phrases of the implicit part.

### 3.2.1 Formation

Let $\mathcal{V}_t$ denote the (denumerable) set of *text identifiers*, $\mathcal{V}_c$ the (denumerable) set of *context identifiers*, and N the positive natural numbers, i.e., $0 \notin$ N. These three sets are assumed to be disjoint. The syntax of the kernel is then specified by the following grammar:

$$
\begin{aligned}
\mathcal{T} \ := \ &\textbf{prim} && (\textit{primitive text}) \\
| \ &\mathcal{V}_t && (\textit{text identifier}) \\
| \ &[\mathcal{C} \vdash \mathcal{T}] && (\textit{abstraction}) \\
| \ &\mathcal{T}(\text{N} := \mathcal{T}) && (\textit{application}) \\
| \ &\mathcal{T} \therefore \mathcal{T} && (\textit{judgment})
\end{aligned}
$$

$$
\begin{aligned}
\mathcal{C} \ := \ &\mathcal{V}_t : \mathcal{T} && (\textit{declaration}) \\
| \ &\mathcal{V}_t := \mathcal{T} && (\textit{definition}) \\
| \ &\textbf{context} \ \mathcal{V}_c := \mathcal{C} && (\textit{context definition}) \\
| \ &\textbf{import} \ \mathcal{V}_c && (\textit{context inclusion}) \\
| \ &[\mathcal{C}; \mathcal{C}] && (\textit{context join})
\end{aligned}
$$

that is the kernel is made up of two syntactic classes, texts denoted by $\mathcal{T}$ and contexts denoted by $\mathcal{C}$.

In the following discussions, let $n$ and $m$ range over N, $x$ and $y$ over $\mathcal{V}_t$, $p$ and $q$ over $\mathcal{V}_c$, $t$ over $\mathcal{T}$, and $c$ over $\mathcal{C}$. Also, indexed and primed variants of these variables shall range over the same sets, e.g. $x_1$ is supposed to range over $\mathcal{V}_t$.

Before proceeding with the formal definitions, we summarize the intended meaning of the language constructs.

### 3.2.2 Intended Meaning of the Constructs

In general, contexts are the means of the Deva language to construct and structure *calculi*, i.e., formalized theories. Among the theories that have been formalized with contexts are: classical and non-classical logics, basic mathematical theories, algebraic data types, and precise methods for software development such as transformation and refinement methods.

A *declaration* $x : t$ declares the identifier $x$ to be of type $t$. As will be seen below, several operations may instantiate this identifier to a text of type $t$. A *definition* $x := t$ defines the identifier $x$ to stand as an abbreviation for the text $t$. A *context definition* **context** $p := c$ defines the identifier $p$ to stand as an abbreviation for the context $c$. The *context inclusion* **import** $p$ of a context identifier $p$ stands for the context $c$ given in a corresponding context definition **context** $p := c$. In a *context join* $[\![\, c_1 ; c_2 \,]\!]$, the declarations and definitions in $c_1$ are made visible in $c_2$. Thus text and context identifiers introduced in $c_1$ may be used in the declarations and definitions of $c_2$ (but not conversely).

The laws defined by contexts may be applied to construct calculations, i.e., derivations, proofs, or developments. Such calculations are called texts.

The *primitive text* **prim** is the starting point for the hierarchical type construction process in Deva. Any applied occurrence $x$ of a *text identifier* must have a corresponding declaration or definition. In an *abstraction* $[c \vdash t]$, the context $c$ can be seen as a set of parameter declarations or a set of assumptions, whereas the text $t$ can be seen as a function body or a conclusion. Thus, the abstraction $[c \vdash t]$ plays the role of a function or an inference rule. An *application* $t_1(n := t_2)$ instantiates the (multi-) function $t_1$ at its $n$-th domain, provided the declared type and the actual type conform. A *judgment* $t_1 \therefore t_2$ is a text $t_1$ together with the assertion that its type is $t_2$.

### 3.2.3 Environments

Declarations, definitions, and context definitions are called *bindings* so that a context can be understood as a structured collection of bindings. In the process of defining semantic notions such as closed-ness, reduction, etc., a frequent operation is to look up the current binding determined by an identifier. Such a current binding can be somewhat complex to compute due to Deva's various context operations (such as inclusion, join, or others to be introduced in later sections). To facilitate the definitions, we will therefore transform a context into a normal form by (partially) evaluating the context operations. The result is a linear list of bindings which reduces the problem of looking up a current binding to a simple search. Such a linear list of bindings is called an *environment*. Each semantic notion will be defined *relative to* an environment. More formally, let $\mathcal{BND}$ denote the subset of $\mathcal{C}$ of bindings, i.e., either a declaration $X : e$, a definition $X := e$, or a context definition **context** $X := e$, and let $\mathcal{ENV}$ denote the data-structure determined by the following operations:

$$\langle \rangle \; : 1 \to \mathcal{ENV} \qquad\qquad (empty\ environment)$$
$$\langle (\cdot) \rangle \; : \mathcal{BND} \to \mathcal{ENV} \qquad\qquad (singleton\ environment)$$
$$\oplus \; : \mathcal{ENV} \times \mathcal{ENV} \to \mathcal{ENV} \qquad\qquad (environment\ concatenation)$$

such that $\oplus$ is associative with the empty environment as identity (i.e., the free monoid over $\mathcal{BND}$). From now on, let $E$ range over environments. To look up the current binding determined by an identifier we can now define a (partial) function operating on environments as follows. Let $X$ range over $\mathcal{V}_t \cup \mathcal{V}_c$ and $e$ range over $\mathcal{T} \cup \mathcal{C}$, and let $X \odot e$ denote a binding then the lookup function is specified by:

$$\langle \rangle \upharpoonright X \quad \text{is undefined,}$$

$$(E \oplus \langle Y \odot e \rangle) \upharpoonright X = \begin{cases} Y \odot e & \text{if } X = Y \\ E \upharpoonright X & \text{otherwise} \end{cases}.$$

Here, $Y$ also ranges over $\mathcal{V}_t \cup \mathcal{V}_c$. So $E \upharpoonright X$ is derived by searching $E$ from right to left for a binding $X \odot e$.

Now, the operation symbol $\oplus$ is overloaded to also account for the operation of "pushing a context onto an environment" by partially evaluating the context operations. The only operation on contexts introduced so far are the three different binding operations, context inclusion, and context join. In later sections, when additional context operations are introduced, we will extend the definition of $\oplus$ accordingly. For the time being, the partial evaluation is specified by

$$E \oplus x : t = E \oplus \langle x : t \rangle,$$
$$E \oplus x := t = E \oplus \langle x := t \rangle,$$
$$E \oplus \textbf{context } p := c = E \oplus \langle \textbf{context } p := c \rangle,$$
$$E \oplus \textbf{import } p = E \oplus c, \quad \text{where } E \upharpoonright p = \textbf{context } p := c,$$
$$E \oplus [\![ c_1; c_2 ]\!] = (E \oplus c_1) \oplus c_2.$$

*Remark.* If one carefully looks at these rules, one will notice a naming problem caused by the fourth rule: Straightforward implementation of this rule could lead to name clashes, if, for instance, the imported context $c$ depends on some identifier which is redefined after the definition of $p$. For the moment, we will defer (Sect. 3.2.6) the problems involved with naming and appeal to the intuition of the reader that in any given text or context the names of identifiers can be chosen in such a way that no name clashes occur.

### 3.2.4 Closed Texts and Closed Contexts

The first semantic assertion to be defined is the one of *closed texts* and *closed contexts* relative to an environment. Figuratively speaking, a text or a context is closed relative to an environment if all applied occurrences of text or context identifiers have a corresponding binding in the environment. Other occurrences of identifiers, e.g. $x$ in $x : \textbf{prim}$ do not require a binding. Of course, the environment should be closed too. Thus, the notions of closed environments and of closed texts and contexts are mutually recursively defined by the following set of rules:

$$\frac{}{\vdash_{cl} \langle \rangle} \qquad \frac{E \vdash_{cl} x : t}{\vdash_{cl} E \oplus x : t} \qquad \frac{E \vdash_{cl} x := t}{\vdash_{cl} E \oplus x := t} \qquad \frac{E \vdash_{cl} \textbf{context } p := c}{\vdash_{cl} E \oplus \textbf{context } p := c}$$

$$\frac{E \vdash_{cl} \textbf{import } p}{\vdash_{cl} E \oplus \textbf{import } p} \qquad \frac{E \vdash_{cl} [\![ c_1; c_2 ]\!]}{\vdash_{cl} E \oplus [\![ c_1; c_2 ]\!]}$$

$$\frac{\vdash_{cl} E}{E \vdash_{cl} \textbf{prim}} \qquad \frac{\vdash_{cl} E \quad E \upharpoonright x \text{ is defined}}{E \vdash_{cl} x} \qquad \frac{E \vdash_{cl} c \quad E \oplus c \vdash_{cl} t}{E \vdash_{cl} [c \vdash t]}$$

$$\frac{E \vdash_{cl} t_1 \quad E \vdash_{cl} t_2}{E \vdash_{cl} t_1(n := t_2)} \qquad \frac{E \vdash_{cl} t_1 \quad E \vdash_{cl} t_2}{E \vdash_{cl} t_1 \therefore t_2}$$

$$\frac{E \vdash_{cl} t}{E \vdash_{cl} x : t} \qquad \frac{E \vdash_{cl} t}{E \vdash_{cl} x := t} \qquad \frac{E \vdash_{cl} c}{E \vdash_{cl} \mathbf{context}\ p := c}$$

$$\frac{\vdash_{cl} E \quad E \restriction p \text{ is defined}}{E \vdash_{cl} \mathbf{import}\ p} \qquad \frac{E \vdash_{cl} c_1 \quad E \oplus c_1 \vdash_{cl} c_2}{E \vdash_{cl} [\![ c_1; c_2 ]\!]}$$

Note that since bindings established in the antecedent of an abstraction are visible to the consequence, in the third rule (of the second block of rules) the context $c$ must be pushed onto the environment relative to which it is checked whether $t$ is closed. A similar remark holds for the last rule, where all bindings of $c_1$ are visible to $c_2$.

A text $t$ is closed, denoted by $\vdash_{cl} t$, if it is closed relative to the empty environment; equally, a context $c$ is closed, denoted by $\vdash_{cl} c$, if it is closed relative to the empty environment.

## 3.2.5   Reduction of Texts and Contexts

This section will introduce an equivalence relation on expressions. The intention is to identify those expressions that are syntactically different, although intuitively they have the same constructive content. There are several possibilities to introduce such an equivalence. In this presentation, it is defined in an operational way by defining a process, called *reduction*, to transform expressions into a *normal form*.

The reduction relation is defined as the reflexive and transitive closure of a "single-step reduction." The single-step reduction of Deva is denoted by $\rhd^1$ where reduction takes place relative to an environment to resolve the binding of identifiers, i.e., $E \vdash t_1 \rhd^1 t_2$ asserts that the text $t_1$ reduces in a single step to $t_2$ relative to the environment $E$, and similarly $E \vdash c_1 \rhd^1 c_2$ asserts that the context $c_1$ reduces in a single step to $c_2$ relative to the environment $E$. The single-step reduction of the kernel of Deva is now specified by the following set of rules which is grouped into *axioms* and *structural rules*. Axioms define the elementary reduction steps and the first group concerns text identifiers and abstractions:

$$E \vdash x \rhd^1 t, \quad \text{if } E \restriction x = x := t,$$
$$E \vdash [x := t_1 \vdash t_2] \rhd^1 t_2, \quad \text{if } E \vdash_{cl} t_2,$$
$$E \vdash [\mathbf{context}\ p := c \vdash t] \rhd^1 t, \quad \text{if } E \vdash_{cl} t,$$
$$E \vdash [\![ c_1; c_2 ]\!] \vdash t] \rhd^1 [c_1 \vdash [c_2 \vdash t]].$$

The first axiom describes the unfolding of a definition and the second and third axiom state that a completely unfolded definition may be removed. The fourth axiom describes the sequentialization of the join operation within the antecedent of an abstraction. The next group of axioms concerns applications and judgments:

$$E \vdash [x : t_1 \vdash t_2](1 := t) \vartriangleright [x := t \vdash t_2],$$
$$E \vdash [x : t_1 \vdash t_2](n + 1 := t) \vartriangleright [x : t_1 \vdash t_2(n := t)],$$
$$E \vdash t_1 \therefore t_2 \vartriangleright t_1.$$

The first two axioms describe the instantiation of the $n$-th declaration by a given argument. Note that in contrast to $\lambda$-calculus, no external substitution operation is introduced. Instead, the internal notion of definition is used, leading to these two axioms which resemble a graph reduction step [62]. Also note, that no applicability condition, i.e., that the type of $t$ matches $t_1$, is imposed on the first axiom. Indeed, at the current point of definition no typing constraints are considered. However, typing constraints will be systematically defined and imposed on expressions further below (Sect. 3.2.9). Similarly note, that texts containing incorrect applications, e.g. $E \vdash \mathbf{prim}(2 := t)$, remain irreducible. Again, these and many kinds of errors will be excluded later. There will never be any need to reduce such situations, it is thus sufficient to consider reduction of "well-behaved" situations only. The third axiom describes the detachment of type information from a judgment. This illustrates again, that at this point of the language definition type constraints are of no concern. The following two axioms relate to contexts:

$$E \vdash \mathbf{import}\, p \vartriangleright c, \quad \text{if } E \upharpoonright p = \mathbf{context}\, p := c,$$
$$E \vdash [\![ [\, c_1; c_2 ]; c_3 ]\!] \vartriangleright [\![ c_1; [\, c_2; c_3 ]\!]].$$

The first axiom, similar to the first axiom of the preceding group, describes the unfolding of a context definition and the second axiom describes the transformation of nested contexts into a right-associative form.

Structural rules state that reduction is monotonic with respect to the syntactic structure of texts and contexts:

$$\frac{E \vdash c_1 \vartriangleright c_2}{E \vdash [c_1 \vdash t] \vartriangleright [c_2 \vdash t]} \qquad \frac{E \oplus c \vdash t_1 \vartriangleright t_2}{E \vdash [c \vdash t_1] \vartriangleright [c \vdash t_2]}$$

$$\frac{E \vdash t_1 \vartriangleright t_2}{E \vdash t_1(n := t) \vartriangleright t_2(n := t)} \qquad \frac{E \vdash t_1 \vartriangleright t_2}{E \vdash t(n := t_1) \vartriangleright t(n := t_2)}$$

$$\frac{E \vdash t_1 \vartriangleright t_2}{E \vdash t_1 \therefore t \vartriangleright t_2 \therefore t} \qquad \frac{E \vdash t_1 \vartriangleright t_2}{E \vdash t \therefore t_1 \vartriangleright t \therefore t_2}$$

$$\frac{E \vdash t_1 \vartriangleright t_2}{E \vdash x : t_1 \vartriangleright x : t_2} \qquad \frac{E \vdash t_1 \vartriangleright t_2}{E \vdash x := t_1 \vartriangleright x := t_2}$$

$$\frac{E \vdash c_1 \vartriangleright c_2}{E \vdash \mathbf{context}\, p := c_1 \vartriangleright \mathbf{context}\, p := c_2}$$

$$\frac{E \vdash c_1 \rhd^1 c_2}{E \vdash [\![ c_1; c ]\!] \rhd^1 [\![ c_2; c ]\!]} \qquad \frac{E \oplus c \vdash c_1 \rhd^1 c_2}{E \vdash [\![ c; c_1 ]\!] \rhd^1 [\![ c; c_2 ]\!]}$$

Note again that the visibility rules of Deva enforce that in the premise of the second and of the last rule the context $c$ is pushed onto the environment $E$.

Now, reduction is specified as the reflexive and transitive closure of single-step reduction:

$$\frac{}{E \vdash t \rhd t} \qquad \frac{E \vdash t_1 \rhd^1 t_2}{E \vdash t_1 \rhd t_2} \qquad \frac{E \vdash t_1 \rhd t_2 \quad E \vdash t_2 \rhd t_3}{E \vdash t_1 \rhd t_3}$$

$$\frac{}{E \vdash c \rhd c} \qquad \frac{E \vdash c_1 \rhd^1 c_2}{E \vdash c_1 \rhd c_2} \qquad \frac{E \vdash c_1 \rhd c_2 \quad E \vdash c_2 \rhd c_3}{E \vdash c_1 \rhd c_3}$$

The properties of reduction, which include confluence and strong normalization under certain restrictions, are presented in detail in Sect. 3.7.

### 3.2.6 Conversion

Conversion is defined to be the equivalence generated by reduction and is denoted by $E \vdash t_1 \sim t_2$ and $E \vdash c_1 \sim c_2$:

$$\frac{}{E \vdash t \sim t} \qquad \frac{E \vdash t_1 \rhd t_2}{E \vdash t_1 \sim t_2} \qquad \frac{E \vdash t_1 \sim t_2}{E \vdash t_2 \sim t_1} \qquad \frac{E \vdash t_1 \sim t_2 \quad E \vdash t_2 \sim t_3}{E \vdash t_1 \sim t_3}$$

$$\frac{}{E \vdash c \sim c} \qquad \frac{E \vdash c_1 \rhd c_2}{E \vdash c_1 \sim c_2} \qquad \frac{E \vdash c_1 \sim c_2}{E \vdash c_2 \sim c_1} \qquad \frac{E \vdash c_1 \sim c_2 \quad E \vdash c_2 \sim c_3}{E \vdash c_1 \sim c_3}$$

*Remark.* One should be aware of the fact that the existence of definitions and rules like the second and the third axiom of elementary reduction give rise to the fact that conversion, unlike reduction, does not preserve the closed-ness of a text or a context. For instance, $E \vdash t_1 \sim t_2$ and $E \vdash_{cl} t_1$ do not, in general, imply that $E \vdash_{cl} t_2$: As a simple example take $t_1 = [a : \mathbf{prim} \vdash a]$, $t_2 = [b := c \vdash [a : \mathbf{prim} \vdash a]]$, and $E = \langle \rangle$.

*Remark.* Now that the reader has gained a first impression of some elements of the formal definition of Deva, we will take a closer look at how identifiers are dealt with. Recall that we assumed that — by a suitable naming of identifiers — name clashes can in general be avoided. The problem of how to avoid name clashes in formal systems is of course not new, and in the past various solutions have been proposed. All solutions introduce ways to abstract from concrete identifiers used for bound variables. The most popular approach, perhaps, is the one used in the typed or untyped $\lambda$-calculus, namely $\alpha$-conversion, which allows the consistent renaming of bound variables. All terms which are equivalent wrt. the equivalence induced by $\alpha$-conversion are then assumed to be equal. However, as simple as this solutions sounds, it is difficult to adapt to our needs since, for example, it collides with Deva's own renaming mechanism to be introduced in the next section.

Thus, we turn to a second approach which does not use concrete names at
all but instead replaces each applied occurrence of an identifier by a *pointer* to
its binder. This so called *nameless notation* originated with de Bruijn and the
AUTOMATH project [26]. He proposed to encode the pointer by the number of
bindings between the occurrence of the identifier and its binding. This encoding,
sometimes called *de Bruijn index*, induces a function which translates a *closed*
$\lambda$-term using concrete identifier names into a term in the nameless notation. To
illustrate the effect of such a translation, consider the following example where
the ideas outlined above are adapted to a Deva text.

$$
\begin{array}{ll}
[\ nat\ : \mathbf{prim} & [\ \llcorner\ : \mathbf{prim} \\
;\ zero\ : nat & ;\ \llcorner\ : 1 \\
;\ f\quad : [\![\, x : nat; y : nat ]\!] \vdash nat] \quad\longmapsto & ;\ \llcorner\ : [\![\, \llcorner : 2; \llcorner : 3 ]\!] \vdash 4] \\
\vdash f(1 := zero) & \vdash 1(1 := 2) \\
] & ]
\end{array}
$$

Note that in order to fully abstract from identifier names, those names which
occur at a binding are replaced by a standard identifier '$\llcorner$.' However, this
information is needed, for example to handle Deva's own renaming mechanism.
Thus, we will decompose the translation into two phases: in a first phase, a
function $\tau$ translates the applied occurrences of identifiers to their de Bruijn
indices but leaves occurrences at bindings unchanged. This is illustrated by the
following example:

$$
\begin{array}{ll}
[\ nat\ : \mathbf{prim} & [\ nat\ : \mathbf{prim} \\
;\ zero\ : nat & ;\ zero\ : 1 \\
;\ f\quad : [\![\, x : nat; y : nat ]\!] \vdash nat] \quad\stackrel{\tau}{\longmapsto} & ;\ f\quad : [\![\, x : 2; y : 3 ]\!] \vdash 4] \\
\vdash f(1 := zero) & \vdash 1(1 := 2) \\
] & ]
\end{array}
$$

In effect, $\tau$ constructs a text with a *built-in symbol table*. Note that restricted to
closed texts, $\tau$ is a bijection since the original identifier can be recaptured from
its binding. In a second phase, a function $\kappa$ replaces the remaining identifiers at
bindings by the standard identifier:

$$
\begin{array}{ll}
[\ nat\ : \mathbf{prim} & [\ \llcorner\ : \mathbf{prim} \\
;\ zero\ : 1 & ;\ \llcorner\ : 1 \\
;\ f\quad : [\![\, x : 2; y : 3 ]\!] \vdash 4] \quad\stackrel{\kappa}{\longmapsto} & ;\ \llcorner\ : [\![\, \llcorner : 2; \llcorner : 3 ]\!] \vdash 4] \\
\vdash 1(1 := 2) & \vdash 1(1 := 2) \\
] & ]
\end{array}
$$

Thus, the composition $\kappa \circ \tau$ is indeed equal to the translation described at the
beginning of this remark.

We could have used this refined indexing scheme for the definition of Deva
from the start, but, as already de Bruijn pointed out, this notation "is not easy

to write and easy to read for the human reader." In order to keep this language definition readable, we propose the following compromise. We assume given a definition of the Deva kernel based on the first translation function $\tau$. This is a rather technical exercise and is included — for the special case of a Deva kernel — in the appendix. The definition given in this and the following sections can then be seen as an idealized presentation, i.e. using names instead of pointers.

It is important to understand the consequences of this convention. For example it is not true that

$$\langle\rangle \vdash [x : \mathbf{prim} \vdash x] \sim [y : \mathbf{prim} \vdash y]$$

since this proposition is an idealized presentation of the underlying proposition

$$\langle\rangle \vdash [x : \mathbf{prim} \vdash 1] \sim [y : \mathbf{prim} \vdash 1]$$

in which the name difference between the two declarations does matter. However, it *is* true that

$$\langle\rangle \vdash \kappa([x : \mathbf{prim} \vdash x]) \sim \kappa([y : \mathbf{prim} \vdash y])$$

since this is an idealized presentation of the underlying proposition

$$\langle\rangle \vdash \kappa([x : \mathbf{prim} \vdash 1]) \sim \kappa([y : \mathbf{prim} \vdash 1])$$

which is clearly true (since $\kappa$ replaces both $x$ and $y$ by $\sqcup$).

Having this in mind, we return to the definition of the Deva kernel and introduce a *name irrelevant conversion*, i.e., conversion modulo $\kappa$. To avoid inconsistencies, we restrict this relation to closed texts and contexts:

$$\frac{E \vdash_{cl} t_1 \quad E \vdash_{cl} t_2 \quad \kappa(E) \vdash \kappa(t_1) \sim \kappa(t_2)}{E \vdash t_1 \stackrel{.}{\sim} t_2}$$

$$\frac{E \vdash_{cl} c_1 \quad E \vdash_{cl} c_2 \quad \kappa(E) \vdash \kappa(c_1) \sim \kappa(c_2)}{E \vdash c_1 \stackrel{.}{\sim} c_2}$$

### 3.2.7   Type Assignment of Texts

Besides the operational concept of reduction, the static concept of *type assignment* is the other fundamental semantic concept of Deva. In Deva, texts are typed: the type of a text relative to an environment is again a text describing the logical effect achieved by that text. For example, the type of a text representing a logical derivation is the theorem established by that derivation.

The function typ : $\mathcal{ENV} \times \mathcal{T} \rightarrow \mathcal{T}$ computes the type of a text and is inductively defined by the following rules. First, the type of the constant **prim** remains undefined:

$$\mathrm{typ}_E(\mathbf{prim}) \text{ is undefined.}$$

The typing rule for an applied occurrence of identifiers is straightforward, i.e., in case the identifier is defined in the environment, the type of a declared identifier is the declared type and the type of a defined identifier is the type of the defining text:

$$\text{typ}_E(x) = \begin{cases} t & \text{if } E \restriction x = x : t \\ \text{typ}_E(t) & \text{if } E \restriction x = x := t. \end{cases}$$

The typing rule for the judgment is also straightforward since the associated type information is of no concern:

$$\text{typ}_E(t_1 \therefore t_2) = \text{typ}_E(t_1).$$

The typing rules for abstraction and application may be motivated as follows: An abstraction can be thought of as an infinite tuple of texts indexed by the declared variables of the abstracted context. For example, the abstraction $[x : t \vdash s(1 := x)]$, where $x$ is assumed not to occur free in $s$, may be visualized as the infinite tuple

$$\langle t' :: \text{typ}(t') = t :: s(1 := t') \rangle.$$

This tuple consists of all terms $s(1 := t')$ such that $\text{typ}(t') = t$. The tuple is also indexed by terms $t'$ such that $\text{typ}(t') = t$. Likewise, an application may be viewed as a projection from such an infinite tuple. It is now quite natural to require that the type of a *finite* tuple of texts is the tuple consisting of the types of the texts, and that the type of a projection from some tuple is the corresponding projection of the type of that tuple. — By the way, this will be the typing rule for products which are introduced in the next section. — It is reasonable to extend this definition to infinite tuples, so that the type of the above tuple is

$$\langle t' :: \text{typ}(t') = t :: \text{typ}(s(1 := t')) \rangle,$$

which we shall view as the type of the abstraction $[x : t \vdash s(1 := x)]$. Accordingly, the type of an application $[x : t \vdash s(1 := x)](1 := t')$ should be $\text{typ}(s(1 := t'))$. This motivates the typing rule for applications:

$$\text{typ}_E(t_1(n := t_2)) = \text{typ}_E(t_1)(n := t_2).$$

Given this rule, the infinite tuple above is equal to

$$\langle t' :: \text{typ}(t') = t :: \text{typ}(s)(1 := t') \rangle,$$

which models the abstraction $[x : t \vdash \text{typ}(s)(1 := x)]$. This motivates the typing rule for abstractions.

$$\text{typ}_E([c \vdash t]) = [c \vdash \text{typ}_{E \oplus c}(t)].$$

A simple example may help to provide some intuition for the typing of abstractions and applications. Assume the environment $E = \langle nat : \mathbf{prim}, o : nat \rangle$ introducing a type $nat$ with an element $o$ and consider the application $[x : nat \vdash x](1 := o)$. While intuitively one would expect it to be of type $nat$, according to the formal typing rules its type is not $nat$ but $[x : nat \vdash nat](1 := o)$. However, it is not difficult to see that, based on the axioms of the previous section, this application is convertible with (and even reducible to) the expected type $nat$.

To summarize, we have the following inductive definition of the typing rules for the Deva kernel.

$$\mathrm{typ}_E(\mathbf{prim}) = \text{undefined},$$

$$\mathrm{typ}_E(x) = \begin{cases} t & \text{if } E \upharpoonright x = x : t \\ \mathrm{typ}_E(t) & \text{if } E \upharpoonright x = x := t \\ \text{undefined} & \text{if } E \upharpoonright x \text{ is undefined} \end{cases},$$

$$\mathrm{typ}_E([c \vdash t]) = [c \vdash \mathrm{typ}_{E \oplus c}(t)],$$

$$\mathrm{typ}_E(t_1(n := t_2)) = \mathrm{typ}_E(t_1)(n := t_2),$$

$$\mathrm{typ}_E(t_1 \therefore t_2) = \mathrm{typ}_E(t_1).$$

*Remark.* Note that, in Deva texts, their types, the types of their types etc. are not given in separate notations but are syntactically identical. This is justified since the given operations of typing and reduction have the same properties on all these levels, thus suggesting introducing a single common data type for these entities. While this leads to a simple syntactic system, the role of a text, i.e., whether it is an element, type, type of type etc., must be determined from the context.

*Remark.* The reader might wonder about the typing rules for contexts. In fact, no useful concept of context typing, i.e., theory typing, could be identified within the technical framework of the Deva language. A framework which allows to express typable theories is presented in [72].

### 3.2.8  Auxiliary Predicates for Validity

In the next section, the notion of validity is defined: a text or a context is valid if certain type constraints (applicability conditions) are satisfied. To specify these type constraints, some auxiliary predicates are needed, namely the family of semantic relations 'domain'. Figuratively speaking, a text $t_1$ is an $n$-th domain of a text $t_2$ relative to an environment $E$, denoted by $n\text{-dom}_E(t_2, t_1)$, if the $n$-th "argument position" of $t_2$ has a type convertible to $t_1$. This notion is made precise by the following rules:

$$\frac{E \vdash \mathrm{typ}_E(t_1) \sim [x : t_2 \vdash t_3]}{1\text{-dom}_E(t_1, t_2)}$$

$$\frac{E \vdash \mathrm{typ}_E(t_1) \sim [x : t_2 \vdash t_3] \qquad n\text{-dom}_{E \oplus x : t_2}(t_1(1 := x), t_4) \qquad E \vdash_{cl} t_4}{(n+1)\text{-dom}_E(t_1, t_4)}$$

$$\frac{E \vdash t_1 \sim [x : t_2 \vdash t_3]}{1\text{-dom}_E(t_1, t_2)}$$

$$\frac{E \vdash t_1 \sim [x : t_2 \vdash t_3] \qquad n\text{-dom}_{E \oplus x : t_2}(t_3, t_4) \qquad E \vdash_{cl} t_4}{(n+1)\text{-dom}_E(t_1, t_4)}$$

*Remark.* One might wonder why it is necessary to have the second pair of rules in addition to the first pair. This is due to the fact that a text like $[x : \mathbf{prim} \vdash \mathbf{prim}]$ is perfectly valid but cannot be typed. Thus its first domain can be determined by the second rule but not by the first. Note that this introduces the possibility that the first domain of texts like $[x : \mathbf{prim} \vdash x]$ can be determined in two different ways. However, the domains determined are unique up to name-free conversion, i.e., it can be shown in general that $n\text{-dom}_E(t, t_1)$ and $n\text{-dom}_E(t, t_2)$ together imply that $E \vdash t_1 \overset{\cdot}{\sim} t_2$. The third premise in the two rules on the right-hand side is necessary, since domains which depend on unknown arguments make no sense, as for instance in the case of the second domain of the text $[x : \mathbf{prim}; y : x \vdash y]$. Note that in the third rule, $t_1$ is applied to an abstract argument $x$ in order to get hold of its subsequent domains. In the last rule this application could be simplified thanks to the property $E \vdash t_1(1 := x) \overset{\cdot}{\sim} t_3$.

### 3.2.9 Validity

Now we are ready to define the central semantic assertion of validity. A text $t$ or a context $c$ is *valid* if every application and every judgment contained in $t$ or $c$ is well-typed. For example, for a text $t$ describing a logical derivation to be valid under an environment $E$ means that the derivation is without gaps and makes proper use of the rules and axioms available in $E$.

This assertion is denoted by $E \vdash_{val} t$ or $E \vdash_{val} c$ and again we require that the environment $E$ is valid too. The rules specifying valid environments are straightforward and moreover identical to those specifying closed environments (Sect. 3.2.4).

$$\frac{}{\vdash_{val} \langle\rangle} \qquad \frac{E \vdash_{val} x : t}{\vdash_{val} E \oplus x : t} \qquad \frac{E \vdash_{val} x := t}{\vdash_{val} E \oplus x := t} \qquad \frac{E \vdash_{val} \mathbf{context}\, p := c}{\vdash_{val} E \oplus \mathbf{context}\, p := c}$$

$$\frac{E \vdash_{val} \mathbf{import}\, p}{\vdash_{val} E \oplus \mathbf{import}\, p} \qquad \frac{E \vdash_{val} [\![ c_1; c_2 ]\!]}{\vdash_{val} E \oplus [\![ c_1; c_2 ]\!]}$$

The interesting rules are those specifying valid texts. Naturally, the primitive text and all bound identifiers are valid. An abstraction is valid if the abstracted context is valid and the text body is valid under the abstracted context. An application $t_1(n := t_2)$ is valid, if the type of the "argument" $t_2$ is an $n$-th domain of the "function" $t_1$. A judgment $t_1 \therefore t_2$ is valid, if the type of the text $t_1$ is name-free convertible to the "type tag" $t_2$.

$$\frac{\vdash_{val} E}{E \vdash_{val} \mathbf{prim}} \qquad \frac{\vdash_{val} E \quad E \restriction x \text{ is defined}}{E \vdash_{val} x} \qquad \frac{E \vdash_{val} c \quad E \oplus c \vdash_{val} t}{E \vdash_{val} [c \vdash t]}$$

$$\frac{E \vdash_{val} t_1 \quad E \vdash_{val} t_2 \quad n\text{-dom}_E(t_1, \mathrm{typ}_E(t_2))}{E \vdash_{val} t_1(n := t_2)}$$

$$\frac{E \vdash_{val} t_1 \quad E \vdash_{val} t_2 \quad E \vdash \mathrm{typ}_E(t_1) \overset{\cdot}{\sim} t_2}{E \vdash_{val} t_1 \therefore t_2}$$

The rules specifying valid contexts are again identical to those specifying closed contexts.

$$\frac{E \vdash_{val} t}{E \vdash_{val} x : t} \qquad \frac{E \vdash_{val} t}{E \vdash_{val} x := t} \qquad \frac{E \vdash_{val} c}{E \vdash_{val} \textbf{context } p := c}$$

$$\frac{\vdash_{val} E \quad E \restriction p \text{ is defined}}{E \vdash_{val} \textbf{import } p} \qquad \frac{E \vdash_{val} c_1 \quad E \oplus c_1 \vdash_{val} c_2}{E \vdash_{val} [\![ c_1; c_2 ]\!]}$$

A text $t$ is valid, denoted by $\vdash_{val} t$, if it is valid relative to the empty environment; equally, a context $c$ is valid, denoted by $\vdash_{val} c$, if it is valid relative to the empty environment.

*Remark.* Note that the structure of the above rules implies that $\vdash_{val}$ is stronger than $\vdash_{cl}$, i.e., whenever $E \vdash_{val} t$ or $E \vdash_{val} c$ holds then $E \vdash_{cl} t$ or $E \vdash_{cl} c$ holds, respectively.

## 3.3   The Explicit Part: Extensions

### 3.3.1   Product

*Formation.* Texts are extended as follows:

$$\mathcal{T} := \overbrace{\langle\!\langle \mathcal{V}_t := \mathcal{T}, \ldots, \mathcal{V}_t := \mathcal{T} \rangle\!\rangle}^{k}, \quad \text{where } k \in \mathbb{N} \qquad (\textit{named product})$$
$$\quad | \ \ \mathcal{T}.\mathbb{N} \qquad\qquad\qquad\qquad\qquad\qquad\qquad (\textit{projection})$$
$$\quad | \ \ \mathcal{T} \textbf{ where } \mathbb{N} := \mathcal{T} \qquad\qquad\qquad\qquad (\textit{modification})$$

Note that there exists a product, selection, and update construction for *each* positive $k$. Thus, the rules presented below really denote *families* of rules indexed by the positive naturals. To this end, let $k$ and $l$ range over $\mathbb{N}$.

*Intended Meaning.* A named product introduces a finite list of named texts. Projection allows to specialize a product to a particular instance. Product modification modifies a component of a product.

*Closed Products.* A product is closed if its components are closed. Likewise, a projection or a modification is closed, if its constituents are closed. This is expressed by the following obvious rules which augment the ones given in Sect. 3.2.4.

$$\frac{E \vdash_{cl} t_1 \quad \cdots \quad E \vdash_{cl} t_k}{E \vdash_{cl} \langle\!\langle x_1 := t_1, \ldots, x_k := t_k \rangle\!\rangle} \qquad \frac{E \vdash_{cl} t}{E \vdash_{cl} t.k} \qquad \frac{E \vdash_{cl} t_1 \quad E \vdash_{cl} t_2}{E \vdash_{cl} t_1 \textbf{ where } k := t_2}$$

*Reduction.* The axioms for *reduction* are extended by the following two axioms which reflect the semantics implied by the notation:

$$E \vdash \langle x_1 := t_1, \ldots, x_k := t_k \rangle.l \; \rhd^1 \; t_l$$

$$E \vdash \langle x_1 := t_1, \ldots, x_k := t_k \rangle \text{ where } l := t$$
$$\rhd^1 \; \langle x_1 := t_1, \ldots, x_l := t, \ldots, x_k := t_k \rangle$$

for $1 \le l \le k$. That is, the projection of a product onto a coordinate reduces to the text at that coordinate and the modification of a product reduces to the modified product. The structural rules are extended in the following evident way:

$$\frac{E \vdash \; t_i \; \rhd^1 \; t_i'}{\begin{array}{c} E \vdash \langle x_1 := t_1, \ldots, x_i := t_i, \ldots, x_k := t_k \rangle \\ \rhd^1 \langle x_1 := t_1, \ldots, x_i := t_i', \ldots, t_k := t_k \rangle \end{array}} \quad , \qquad \text{for } 1 \le i \le k$$

$$\frac{E \vdash \; t_1 \; \rhd^1 \; t_2}{E \vdash \; t_1.k \; \rhd^1 \; t_2.k} \qquad \frac{E \vdash \; t_1 \; \rhd^1 \; t_1'}{E \vdash \; t_1 \text{ where } k := t \; \rhd^1 \; t_1' \text{ where } k := t}$$

$$\frac{E \vdash \; t_1 \; \rhd^1 \; t_1'}{E \vdash \; t \text{ where } k := t_1 \; \rhd^1 \; t \text{ where } k := t_1'}$$

*Type Assignment.* As was already pointed out in Sect. 3.2.7, the type of a finite product is quite naturally defined to be the product of the types of the components

$$\mathrm{typ}_E(\langle x_1 := t_1, \ldots, x_k := t_k \rangle) = \langle x_1 := \mathrm{typ}_E(t_1), \ldots, x_k := \mathrm{typ}_E(t_k) \rangle.$$

This rule implies that the type of the projection from some product is the corresponding projection from the type of the product

$$\mathrm{typ}_E(t.k) = \mathrm{typ}_E(t).k,$$

and that the type of the modification of a product by some text is the modification of the type of the product by the type of the text

$$\mathrm{typ}_E(t_1 \text{ where } k := t_2) = \mathrm{typ}_E(t_1) \text{ where } k := \mathrm{typ}_E(t_2).$$

*Validity.* The rules for validity are extended as follows: A product is valid if its components are valid and if their names are disjoint:

$$\frac{E \vdash_{val} t_1 \quad \cdots \quad E \vdash_{val} t_k \qquad x_i \ne x_j \text{ for } 1 \le i < j \le k}{E \vdash_{val} \langle x_1 := t_1, \ldots, x_k := t_k \rangle} \; .$$

A projection of some text $t$ onto a coordinate $l$ is valid if $t$ is convertible (modulo renaming) to some product with $k$ components and $1 \le l \le k$:

$$\frac{E \vdash_{val} t \quad 1 \le l \le k}{E \vdash_{val} t \sim \langle x_1 := t_1, \ldots, x_k := t_k \rangle} \qquad \frac{E \vdash_{val} t \quad 1 \le l \le k}{E \vdash_{val} \mathrm{typ}_E(t) \sim \langle x_1 := t_1, \ldots, x_k := t_k \rangle}$$
$$\frac{}{E \vdash_{val} t.l} \qquad\qquad\qquad \frac{}{E \vdash_{val} t.l}$$

The validity of a modification is defined similarly:

$$\frac{E \vdash_{val} t \quad E \vdash_{val} t' \quad 1 \leq l \leq k}{E \vdash_{val} t \stackrel{\cdot}{\sim} [x_1 := t_1, \ldots, x_k := t_k]} \qquad \frac{E \vdash_{val} t \quad E \vdash_{val} t' \quad 1 \leq l \leq k}{E \vdash_{val} \mathrm{typ}_E(t) \stackrel{\cdot}{\sim} \langle x_1 := t_1, \ldots, x_k := t_k \rangle}$$
$$E \vdash_{val} t \text{ where } l := t' \qquad\qquad\qquad E \vdash_{val} t \text{ where } l := t'$$

In particular, it is not excluded that the modification of a product component alters its type.

*Remark.* In the two preceding groups of rules, the first rule is necessary since **prim** is a valid component of a product which cannot be typed. On the other hand, the second rule is necessary, since a projection of or a modification from a text whose type is a product but which itself is not should also be valid.

### 3.3.2   Sum

*Formation.* Texts are extended as follows:

$$\mathcal{T} := \overbrace{\langle \mathcal{V}_t := \hat{\ }^r \mathcal{T} \mid \cdots \mid \mathcal{V}_t := \hat{\ }^r \mathcal{T} \mid \mathcal{V}_t := \mathcal{T} \mid \mathcal{V}_t := \hat{\ }^r \mathcal{T} \mid \cdots \mid \mathcal{V}_t := \hat{\ }^r \mathcal{T} \rangle}^{k}$$
$$\text{where } k \in \mathbb{N} \text{ and } r \in \mathbb{N} \cup \{0\}, \qquad\qquad (\textit{named sum})$$
$$\mid \textbf{case } \mathcal{T} \textbf{ of } \mathcal{T} \qquad\qquad\qquad (\textit{case distinction})$$

That is, a named sum consists of a positive number of components, at least one of which has the form $\mathcal{V}_t := \mathcal{T}$ and, for some $r \in \mathbb{N} \cup \{0\}$, the remaining ones have the form $\mathcal{V}_t := \hat{\ }^r \mathcal{T}$. Here, $\hat{\ }^r$ denotes $r$ times the application of the *ghost* operator $\hat{\ }$. Thus $\hat{\ }^0 t$ denotes $t$. Since $r$ is unique we are justified to write

$$\langle x_1 := \hat{\ }^r t_1 \mid \cdots \mid x_i := t_i \mid \cdots \mid x_k := \hat{\ }^r t_k \rangle$$

for the more elaborate

$$\langle x_1 := \hat{\ }^r t_1 \mid \cdots \mid x_{i-1} := \hat{\ }^r t_{i-1} \mid x_i := t_i \mid x_{i+1} := \hat{\ }^r t_{i+1} \mid \cdots \mid x_k := \hat{\ }^r t_k \rangle,$$

which helps to avoid a lot of notational clutter.

*Intended Meaning.* In addition to the named product, the named sum is another construct to collect a finite list of texts. Sums come in several variants: We will call a sum like the one above an $r$–*injection*. In the special case where $r$ is zero, we also use the term *plain sum*. One can think of a plain sum $\langle x_1 := t_1 \mid x_2 := t_2 \rangle$ to model the disjoint union, or variant record, of all texts whose type is equal to $t_1$ and all texts whose type is equal to $t_2$. If the type of, say, $t_1'$ is convertible to $t_1$ the 1-injection $\langle x_1 := t_1' \mid x_2 := \hat{\ } t_2 \rangle$ then stands for the tagged element $t_1'$ of $\langle x_1 := t_1 \mid x_2 := t_2 \rangle$. As could be seen in the previous section, typing in Deva is defined as a homomorphic mapping from texts into texts. As a consequence $r$-th types of texts can be constructed by repeatedly applying the type mapping $r$ times to texts. This allows to introduce generalized $r$–*injections*, behaving as 1-injections, except that the $r$-th type of the unrealized variants is specified. As for products, we assume in the following that $r$ ranges over the naturals, and that $k$ and $l$ range over the positive naturals. In general, a sum has properties weaker than those of the product, i.e., projection is not allowed. Instead, by a case distinction **case** $t_1$ **of** $t_2$ one may select from a number of alternatives functions in $t_2$, depending on the kind of injection applied to the argument $t_1$. Sums control the static description of these alternatives as well as the selection process.

*Closed Sums.* Similar to products, a named sum or a case distinction is closed if its components are closed. This is expressed by the following two rules which augment the ones already given.

$$\frac{E \vdash_{cl} t_1 \quad \cdots \quad E \vdash_{cl} t_k}{E \vdash_{cl} \langle x_1 := {}^{\frown^r} t_1 \mid \cdots \mid x_i := t_i \mid \cdots \mid x_k := {}^{\frown^r} t_k \rangle}$$

$$\frac{E \vdash_{cl} t \quad E \vdash_{cl} t'}{E \vdash_{cl} \textbf{case } t \textbf{ of } t'}$$

*Reduction.* The intended use of sums becomes evident by the following axiom of reduction:

$$E \vdash \textbf{case} \langle x_1 := {}^{\frown} t_1 \mid \cdots \mid x_i := t_i \mid \cdots \mid x_k := {}^{\frown} t_k \rangle \textbf{ of } \langle y_1 := t_1', \dots, y_k := t_k' \rangle$$
$$\rhd^1 t_i'(1 := t_i)$$

This axiom describes the selection of a product component by means of 1–injections.

As always, the additional structural rules are evident $(1 \le i \le k)$:

$$\frac{E \vdash t_i \rhd^1 t_i'}{\begin{array}{c} E \vdash \langle x_1 := {}^{\frown^r} t_1 \mid \cdots \mid x_i := t_i \mid \cdots \mid x_k := {}^{\frown^r} t_k \rangle \\ \rhd^1 \langle x_1 := {}^{\frown^r} t_1 \mid \cdots \mid x_i := t_i' \mid \cdots \mid x_k := {}^{\frown^r} t_k \rangle \end{array}}$$

$$\frac{E \vdash t_1 \rhd^1 t_2}{E \vdash \textbf{case } t_1 \textbf{ of } t \rhd^1 \textbf{case } t_2 \textbf{ of } t} \qquad \frac{E \vdash t_1 \rhd^1 t_2}{E \vdash \textbf{case } t \textbf{ of } t_1 \rhd^1 \textbf{case } t \textbf{ of } t_2}$$

*Type Assignment.* The typing rules for sums are now a straightforward consequence of the introductory remarks: The type of a plain sum is the sum of the types of the components

$$\text{typ}_E(\langle x_1 := t_1 \mid \cdots \mid x_k := t_k \rangle) = \langle x_1 := \text{typ}_E(t_1) \mid \cdots \mid x_k := \text{typ}_E(t_k) \rangle.$$

The type of a 1–injection should naturally be the sum into which is injected, thus

$$\text{typ}_E(\langle x_1 := {}^{\frown} t_1 \mid \cdots \mid x_i := t_i \mid \cdots \mid x_k := {}^{\frown} t_k \rangle)$$
$$= \langle [x_1 := t_1 \mid \cdots \mid x_i := \text{typ}_E(t_i) \mid \cdots \mid x_k := t_k \rangle.$$

This rule extends to any positive $r$ as follows

$$\text{typ}_E(\langle x_1 := {}^{\cdot r+1} t_1 \mid \cdots \mid x_i := t_i \mid \cdots \mid x_k := {}^{\cdot r+1} t_k \rangle)$$
$$= \langle x_1 := {}^{\cdot r} t_1 \mid \cdots \mid x_i := \text{typ}_E(t_i) \mid \cdots \mid x_k := {}^{\cdot r} t_k \rangle.$$

Finally, the type of a case distinction is derived from the corresponding reduction rule:

$$\text{typ}_E(\textbf{case } t_1 \textbf{ of } t_2) = \textbf{case } t_1 \textbf{ of } \text{typ}_E(t_2).$$

*Validity.* As for products, a sum is valid if its components are valid

$$\frac{E \vdash_{val} t_1 \quad \cdots \quad E \vdash_{val} t_k}{E \vdash_{val} \langle x_1 := {}^{\cdot r} t_1 \mid \cdots \mid x_i := t_i \mid \cdots \mid x_k := {}^{\cdot r} t_k \rangle}.$$

The only interesting rule is the one defining whether a case distinction is valid or not. If one looks at the reduction axiom for case distinctions, it should be clear that a case distinction **case** $t$ **of** $t'$ makes sense only if $t'$ is a product of abstractions $\langle y_1 := [x_1 : t_1 \vdash t'_1], \ldots, y_k := [x_k : t_k \vdash t'_k] \rangle$ and $t$ is a 1–injection into the plain sum $\langle x_1 := t_1 \mid \cdots \mid x_k := t_k \rangle$. This can succinctly be stated by saying that the type of $t$ is a *domain–sum* of $t'$. To formulate the rules for validity, we therefore define the predicate dom-sum first:

$$\frac{E \vdash \ t \sim \langle y_1 := [x_1 : t_1 \vdash t'_1], \ldots, y_k := [x_k : t_k \vdash t'_k] \rangle}{\text{dom-sum}_E(t, \langle x_1 := t_1 \mid \cdots \mid x_k := t_k \rangle)}$$

$$\frac{E \vdash \ \text{typ}_E(t) \sim \langle y_1 := [x_1 : t_1 \vdash t'_1], \ldots, y_k := [x_k : t_k \vdash t'_k] \rangle}{\text{dom-sum}_E(t, \langle x_1 := t_1 \mid \cdots \mid x_k := t_k \rangle)}$$

With this predicate available, the definition of the validity–rule for case distinctions turns out to be very simple

$$\frac{E \vdash_{val} t \quad E \vdash_{val} t' \quad \text{dom-sum}_E(t', \text{typ}_E(t))}{E \vdash_{val} \textbf{case } t \textbf{ of } t'}$$

### 3.3.3   Cut

*Formation.* Texts are extended as follows:

$$\mathcal{T} := \mathcal{T} \textbf{ upto } N \diamond \mathcal{T} \textbf{ at } N \qquad\qquad (cut)$$

*Intended Meaning.* Whereas application is a point level operation between, i.e., a text acting as a function applied to a text acting as an argument, cuts introduce a way to compose texts both acting as functions.

For example, the text

$$[x : t \vdash f(1 := x)] \textbf{ upto } 1 \Leftrightarrow [y : t' \vdash g(1 := y)] \textbf{ at } 1$$

reduces to the text $[\![\, x : t; y := f(1 := x)]\!] \vdash g(1 := y)]$ which further reduces to $[x : t \vdash g(1 := f(1 := x))]$. The general case simply allows to specify the concrete argument positions at which the composition shall take place. The name "cut" has been chosen, because what happens at the type level of a cut is very akin to the cut-rule in natural deduction, for example the reduction of a cut resembles cut-elimination.

*Closed Cuts.* A cut is closed if its components are closed.

$$\frac{E \vdash_{cl} t \quad E \vdash_{cl} t'}{E \vdash_{cl} t \textbf{ upto } m \Leftrightarrow t' \textbf{ at } n}$$

*Reduction.* The reduction rules express the intended compositional nature of cuts.

$$E \vdash [x : t_1 \vdash t_2] \textbf{ upto } 1 \Leftrightarrow [y : t_3 \vdash t_4] \textbf{ at } 1 \; \rhd \; [\![\, x : t_1; y := t_2]\!] \vdash t_4]$$
$$E \vdash t \textbf{ upto } m \Leftrightarrow [y : t_1 \vdash t_2] \textbf{ at } n+1 \; \rhd \; [y : t_1 \vdash t \textbf{ upto } m \Leftrightarrow t_2 \textbf{ at } n]$$
$$E \vdash [x : t_1 \vdash t_2] \textbf{ upto } m+1 \Leftrightarrow t \textbf{ at } 1 \; \rhd \; [x : t_1 \vdash t_2 \textbf{ upto } m \Leftrightarrow t \textbf{ at } 1]$$

Note again, that type constraints are of no concern at this point of definition. The last two rules describe a structural extension of the cut in a similar way as application was extended above. Note that in the general case, $t \textbf{ upto } m \Leftrightarrow t' \textbf{ at } n$, these rules enforce that the arguments of $t'$ are processed first and the arguments of $t$ last.

Again, the additional structural rules are evident

$$\frac{E \vdash t_1 \rhd^1 t_2}{E \vdash t_1 \textbf{ upto } m \Leftrightarrow t \textbf{ at } n \rhd^1 t_2 \textbf{ upto } m \Leftrightarrow t \textbf{ at } n}$$

$$\frac{E \vdash t_1 \rhd^1 t_2}{E \vdash t \textbf{ upto } m \Leftrightarrow t_1 \textbf{ at } n \rhd^1 t \textbf{ upto } m \Leftrightarrow t_2 \textbf{ at } n}$$

*Type Assignment.* The typing rule for cuts is easily derived from the reduction rules above and the reduction rule for application.

$$\text{typ}_E(t \textbf{ upto } m \Leftrightarrow t' \textbf{ at } n) = t \textbf{ upto } m \Leftrightarrow \text{typ}_E(t') \textbf{ at } n$$

*Validity.* Looking at the reduction rules for cuts, one easily recognizes that a cut $t$ **upto** $m \Diamond t'$ **at** $n$ is valid if $\mathrm{typ}(t)$ is convertible to a form $[\![\, x_1 : t_1; \dots; x_m : t_m ]\!] \vdash t_{m+1}]$, $t'$ is convertible to a form $[\![\, x'_1 : t'_1; \dots; x'_n : t'_n ]\!] \vdash t'_{n+1}]$, and the type of $t_{m+1}$ is convertible to $t'_n$, where of course both $t_{m+1}$ and $t'_n$ have to be closed under $E$. As for sums, this validity rule is more readily formulated with the help of two auxiliary predicates. The first one is already available, namely the predicate $n$-dom of Sect. 3.2.8. The second predicate, $n$-ran, determines whether a text is an $n$-*th range* of another text. It is defined as follows

$$\frac{E \vdash \mathrm{typ}_E(t) \sim [x : t_1 \vdash t_2] \quad E \vdash_{cl} t_2}{1\text{-ran}_E(t, t_2)}$$

$$\frac{E \vdash \mathrm{typ}_E(t) \sim [x : t_1 \vdash t_2]}{n\text{-ran}_{E \oplus x : t_1}(t(1 := x), t_3) \quad E \vdash_{cl} t_3}{n + 1\text{-ran}_E(t, t_3)}$$

Now the validity–rule for cuts is easy to express

$$\frac{E \vdash_{val} t_1 \quad E \vdash_{val} t_2 \quad n\text{-dom}_E(t_2, t) \quad m\text{-ran}_E(t_1, t)}{E \vdash_{val} t_1 \ \textbf{upto} \ m \Diamond t_2 \ \textbf{at} \ n}$$

### 3.3.4   Context Operations

*Formation.* In this section we will introduce some operations on contexts. In particular, *context application* allows the instantiation of a declared variable to some text, and *context renaming* permits the renaming of text or context variables. We will start with extending the syntax for contexts:

$$\begin{aligned}
\mathcal{C} \ &:= \mathcal{C}(\mathrm{N} := \mathcal{T}) && (\textit{context application})\\
&\mid \ \mathcal{C}[\mathcal{V}_t =: \mathcal{V}_t]\\
&\mid \ \mathcal{C}[\mathcal{V}_c =: \mathcal{V}_c] && (\textit{context renaming})
\end{aligned}$$

*Intended Meaning.* A context application $c(n := t)$ instantiates the $n$-th declared identifier in $c$ to the text $t$, provided the declared type and the actual type conform. A *renaming* $c[x =: y]$ renames the text identifier $x$ to $y$ in $c$. A *renaming* $c[p =: q]$ renames the context identifier $p$ to $q$ in $c$.

*Environments.* These three operations on contexts make it necessary to adapt the rules for pushing a context onto an environment which were designed to partially evaluate the context operations (cf. Sect. 3.2.3). First, we give the rules for evaluating a context application: a context application $c(n := t)$ instantiates the $n$–th declared variable in $c$ to $t$. This is expressed by the following four rules

$$\frac{E \oplus c = E \oplus \langle x : t \rangle \oplus E'}{E \oplus c(1 := t') = E \oplus \langle x := t' \rangle \oplus E'}$$

$$\frac{E \oplus c = E \oplus \langle x : t \rangle \oplus c'}{E \oplus c(n + 1 := t') = E \oplus \langle x : t \rangle \oplus c'(n := t')}$$

$$\frac{E \oplus c = E \oplus \langle x := t \rangle \oplus c'}{E \oplus c(n := t') = E \oplus \langle x := t \rangle \oplus c'(n := t')}$$

$$\frac{E \oplus c = E \oplus \langle \mathbf{context}\ p := c_2 \rangle \oplus c_1}{E \oplus c(n := t) = E \oplus \langle \mathbf{context}\ p := c_2 \rangle \oplus c_1(n := t)}$$

Note that the first rule does not enforce any type constraints on $t$ since this is of no concern at this stage. However, the validity rules for context applications will of course do so. Next, we give the rules for context renaming:

$$\frac{E \oplus c = E \oplus \langle X \odot e \rangle \oplus E'}{E \oplus c[X =: Y] = E \oplus \langle Y \odot e \rangle \oplus E'}$$

$$\frac{E \oplus c = E \oplus \langle X \odot e \rangle \oplus c' \quad X \neq Y}{E \oplus c[Y =: Z] = E \oplus \langle X \odot e \rangle \oplus c'[Y =: Z]}$$

Remember that this definition is an idealized presentation of a formalized definition using de-Bruijn indices. This means that only defining occurrences of variables have to be taken care of by the above rules and no adjustment of applied occurrences has to be made.

In order to ensure that the renaming operation is defined, we require that the context to which the renaming operation is applied binds the variable to be renamed. This requirement is stated by the auxiliary predicate 'binds' defined as follows

$$\frac{E \oplus c = E \oplus E' \oplus \langle X \odot e \rangle \oplus E''}{\mathrm{binds}_E(c, X)}$$

*Closed Context Operations.* The rules for closed context operations are straightforward and are given below

$$\frac{E \vdash_{cl} c \quad E \vdash_{cl} t}{E \vdash_{cl} c(n := t)} \qquad \frac{E \vdash_{cl} c \quad \mathrm{binds}_E(c, X)}{E \vdash_{cl} c[X =: Y]}$$

*Reduction.* The reduction rules for context application are quite similar to those for text application and reflect the rules for their partial evaluation

$$E \vdash (x : t)(1 := t') \; \triangleright \; x := t'$$
$$E \vdash [\![ x : t; c ]\!](1 := t') \; \triangleright \; [\![ x := t'; c ]\!]$$
$$E \vdash [\![ x : t; c ]\!](n + 1 := t') \; \triangleright \; [\![ x : t; c(n := t') ]\!]$$
$$E \vdash [\![ x := t; c ]\!](n := t') \; \triangleright \; [\![ x := t; c(n := t') ]\!]$$
$$E \vdash [\![ \textbf{context}\, p := c; c' ]\!](n := t) \; \triangleright \; [\![ \textbf{context}\, p := c; c'(n := t') ]\!]$$

Likewise, the reduction rules for context renaming reflect the rules for their partial evaluation

$$E \vdash (X \odot e)[X =: Y] \; \triangleright \; Y \odot e$$
$$E \vdash [\![ (X \odot e); c ]\!][X =: Y] \; \triangleright \; [\![ Y \odot e; c ]\!]$$
$$E \vdash [\![ (X \odot e); c ]\!][Y =: Z] \; \triangleright \; [\![ X \odot e; c[Y =: Z] ]\!], \quad \text{if } X \neq Y$$

Once again, the structural rules have to be extended in the following evident way

$$\frac{E \vdash c \triangleright c'}{E \vdash c(n := t) \triangleright c'(n := t)} \qquad \frac{E \vdash t \triangleright t'}{E \vdash c(n := t) \triangleright c(n := t')}$$

$$\frac{E \vdash c \triangleright c'}{E \vdash c[X =: Y] \triangleright c'[X =: Y]}$$

*Validity.* For the definition of the validity of context application we will need a "domain" predicate similar to the one defined for texts, i.e., a text $t$ is an $n$-th domain of a context $c$ relative to an environment $E$, denoted by $n\text{-dom}_E(c, t)$, if the $n$-th declaration of $c$ has a type convertible to $t$:

$$\frac{E \vdash c \sim x : t}{1\text{-dom}_E(c, t)} \qquad \frac{E \vdash c \sim [\![ x : t; c' ]\!]}{1\text{-dom}_E(c, t)}$$

$$\frac{E \vdash c \sim [\![ x : t; c' ]\!] \qquad E \vdash n\text{-dom}_{E \oplus x : t}(c', t') \quad E \vdash_{cl} t'}{(n + 1)\text{-dom}_E(c, t')}$$

$$\frac{E \vdash c \sim [\![ x := t; c' ]\!] \qquad E \vdash n\text{-dom}_{E \oplus x := t}(c', t') \quad E \vdash_{cl} t'}{n\text{-dom}_E(c, t')}$$

$$\frac{E \vdash c \sim [\![ \textbf{context}\, p := c; c' ]\!] \qquad E \vdash n\text{-dom}_{E \oplus \textbf{context}\, p := c}(c', t') \quad E \vdash_{cl} t'}{n\text{-dom}_E(c, t')}$$

Now it is not hard to state the rules for the validity of the additional context operations

$$\frac{E \vdash_{val} c \quad E \vdash_{val} t \quad n\text{-dom}_E(c, \text{typ}_E(t))}{E \vdash_{val} c(n := t)} \qquad \frac{E \vdash_{val} c \quad \text{binds}_E(c, X)}{E \vdash_{val} c[X =: Y]}$$

## 3.4 The Explicit Part: Illustrations

Fig. 11 and 12 present the examples from Sect. 3.1 in full detail using the constructs from the explicit part of Deva. The reader should compare the explicit versions with their corresponding implicit versions. A precise technical relation between implicit and explicit expressions will be defined in Sect. 3.5. Some syntactic sugar has been used to enhance readability. In Fig. 11, the declaration $(\cdot) \Rightarrow (\cdot)$ allows to apply $\Rightarrow$ in infix notation, i.e., $p \Rightarrow q$ is equivalent syntax for $(\Rightarrow (1 := p))(1 := q)$. This double application describes the instantiation of $\Rightarrow$ to $p$ at its first domain and of $(\Rightarrow (1 := p))$ to $q$ at its first domain (which amounts to instantiating $\Rightarrow$ to $q$ at its second domain). Furthermore, the following equivalent syntax is used:

$$[\![ c_1; c_2; \cdots ; c_{n-1}; c_n ]\!] \simeq [\![ c_1; [\![ c_2; \cdots ; [\![ c_{n-1}; c_n ]\!] \cdots ]\!] ]\!]$$
$$[t_1 \vdash t_2] \simeq [x : t_1 \vdash t_2], \quad x \text{ new}$$
$$[\![ t_1; t_2 ]\!] \vdash t] \simeq [\![ x : t_1; y : t_2 ]\!] \vdash t], \quad x \text{ and } y \text{ new}$$
$$x, y : t \simeq [\![ x : t; y : t ]\!]$$
$$t(1 := t_1, 2 := t_2) \simeq (t(1 := t_1))(1 := t_2)$$
$$[t_1, t_2] \simeq [x := t_1, y := t_2], \quad x \text{ and } y \text{ new}$$
$$[t_1 \mid t_2] \simeq [x := t_1 \mid y := t_2], \quad x \text{ and } y \text{ new}$$

The last convention for sums is used likewise for injections. Note, that the explicit version of minimal logic differs from the implicit one only in the use of colons within declarations (declarations with question-mark will be defined in Sect. 3.5).

The explicit version of the Syllogism proof contains several new applications and abstractions. Nevertheless, the overall structure of the proof has been preserved. In contrast, Fig. 12 is a rather significant expansion of the implicit version of the simplification example of Fig. 10. The following paragraphs will illustrate the semantic notions of the explicit part by using various excerpts from these examples.

*Illustration of environment operations.* Importing the context *MinimalLogic* in Fig. 11 is easily computed as follows:

$$\langle \textbf{context } \mathit{MinimalLogic} := \cdots \rangle \oplus \textbf{import } \mathit{MinimalLogic}$$
$$=$$
$$\langle \textbf{context } \mathit{MinimalLogic} := \cdots \rangle \oplus [\![ \mathit{prop} : \textbf{prim}; \cdots ]\!]$$
$$=$$
$$\langle \textbf{context } \mathit{MinimalLogic} := \cdots \rangle \oplus \langle \mathit{prop} : \textbf{prim} \rangle \oplus [\![ \cdots ]\!]$$

**context** *Syllogism* :=
[ **context** *MinimalLogic* :=
  [ *prop*    : **prim**
  ; $(\cdot) \Rightarrow (\cdot) : [prop; prop \vdash prop]$
  ; *rules*   : ⟨ *intro* := $[p, q : prop \vdash [[p \vdash q] \vdash p \Rightarrow q]]$
          , *elim* := $[p, q : prop \vdash [p \Rightarrow q \vdash [p \vdash q]]]$
          ▶
  ]
; **import** *MinimalLogic*
; *thesis* := $[a, b, c : prop \vdash [a \Rightarrow b; b \Rightarrow c \vdash a \Rightarrow c]]$
; *proof* := $[a, b, c : prop ; asm_1 : a \Rightarrow b; asm_2 : b \Rightarrow c$
        $\vdash rules .\mathbf{1}(\mathbf{1} := a, \mathbf{2} := c)(\mathbf{1} :=$
          $(([\,u : a \vdash ((rules.\mathbf{2}(\mathbf{1} := a, \mathbf{2} := b))(\mathbf{1} := asm_1))(\mathbf{1} := u)\,] \therefore [a \vdash b])$
            **upto 1** ◇
          $([\,v : b \vdash ((rules.\mathbf{2}(\mathbf{1} := b, \mathbf{2} := c))(\mathbf{1} := asm_2))(\mathbf{1} := v)\,] \therefore [b \vdash c])$
            **at 1**)
          $\therefore [a \vdash c]$
        $) \therefore a \Rightarrow c$
    ]
        $\therefore thesis$
]

**Fig. 11.** Explicit version of Fig. 9

*Illustration of reduction.* The first example illustrates the reduction of projections and applications:

$$\begin{array}{l} ⟨\, intro := [p, q : prop \vdash [[p \vdash q] \vdash p \Rightarrow q]] \\ , elim := [p, q : prop \vdash [p \Rightarrow q \vdash [p \vdash q]]] \\ ⟩.2(1 := b, 2 := c) \end{array}$$

▷  { projection }
  $[p, q : prop \vdash [p \Rightarrow q \vdash [p \vdash q]]](1 := b, 2 := c)$
▷  { application }
  $[p := b \vdash [q : prop \vdash [p \Rightarrow q \vdash [p \vdash q]]]](1 := c)$
▷  { unfolding of definitions }
  $[p := b \vdash [q : prop \vdash [b \Rightarrow q \vdash [b \vdash q]]]](1 := c)$
▷  { removal of definition }
  $[q : prop \vdash [b \Rightarrow q \vdash [b \vdash q]]](1 := c)$
▷  { application }

$\text{context } \textit{Simplification} :=$

$[\![\ \textit{bool} \qquad\qquad :\quad \mathbf{prim}$

$;\ \textit{true}, \textit{false} \qquad :\quad \textit{bool}$

$;\ (\cdot) \wedge (\cdot),$

$\quad (\cdot) \vee (\cdot) \qquad :\quad [\![\,\textit{bool}; \textit{bool} \vdash \textit{bool}\,]\!]$

$;\ \textit{laws} \qquad\qquad :\quad \langle\!\langle\ \textit{and} := [\![\ a : \textit{bool}\ \vdash [\![\ a \wedge \textit{true} \vdash a\,]\!]\,]\!]$

$\qquad\qquad\qquad\qquad ,\ \textit{or}\ := [\![\ a : \textit{bool}\ \vdash [\![\ a \vee \textit{false} \vdash a\,]\!]\,]\!]$

$\qquad\qquad\qquad\ \rangle\!\rangle$

$;\ \textit{simplify} \qquad := [\![\ a\ :\ \textit{bool}$

$\qquad\qquad\qquad\qquad \vdash [\![\ u\ :\ \langle\!\langle\ a \wedge \textit{true}\ |\ a \vee \textit{false}\ \rangle\!\rangle$

$\qquad\qquad\qquad\qquad\quad \vdash \mathbf{case}\ u\ \mathbf{of}\ \langle\!\langle\ \textit{laws}\ .1(1 := a),\ \textit{laws}\ .2(1 := a)\rangle\!\rangle$

$\qquad\qquad\qquad\qquad\quad\ ]\!]$

$\qquad\qquad\qquad\quad\ ]\!]$

$;\ \textit{simplification} := [\![\ b \qquad :\qquad \textit{bool}$

$\qquad\qquad\qquad\qquad\ ;\ u\quad :\quad (b \wedge \textit{true}) \vee \textit{false}$

$\qquad\qquad\qquad\qquad\ ;\ \textit{aux} := (\textit{simplify}(1 := (b \wedge \textit{true})))$

$\qquad\qquad\qquad\qquad\qquad\qquad (1 := \langle\!\langle\ \hat{}\,((b \wedge \textit{true}) \wedge \textit{true})|\ u\ \rangle\!\rangle$

$\qquad\qquad\qquad\qquad\qquad )$

$\qquad\qquad\qquad\qquad\ \vdash (\textit{simplify}(1 := b))(1 := \langle\!\langle\ \textit{aux}\ |\ \hat{}\,(b \vee \textit{false})\rangle\!\rangle$

$\qquad\qquad\qquad\qquad\ )\ \therefore\ b$

$\qquad\qquad\qquad\ ]\!]$

$]\!]$

**Fig. 12.** Explicit version of Fig. 10

$[q := c \vdash [b \Rightarrow q \vdash [b \vdash q]]]$

$\rhd\quad \{\ \text{unfolding of definitions}\ \}$

$[q := c \vdash [b \Rightarrow c \vdash [b \vdash c]]]$

$\rhd\quad \{\ \text{removal of definition}\ \}$

$[b \Rightarrow c \vdash [b \vdash c]]$

The environment in which this reduction takes place has been omitted. It is assumed to be as within the proof of Syllogism in Fig. 11. In some of the above reduction steps the structural rules have been tacitly applied. It is interesting to extend this reduction along the lines of the Syllogism proof. Remember that $[t_1 \vdash t_2]$ is equivalent syntax for $[x : t_1 \vdash t_2]$, where $x$ is a new identifier:

$((\langle\!\langle\ \textit{intro}\ := [p, q : \textit{prop} \vdash [[p \vdash q] \vdash p \Rightarrow q]]$

$\quad ,\ \textit{elim}\ := [p, q : \textit{prop} \vdash [p \Rightarrow q \vdash [p \vdash q]]]$

$\rangle\!\rangle.2(1 := b, 2 := c))(1 := \textit{hyp}_2, 2 := v)$

$\rhd\quad \{\ \text{above, structural rules}\ \}$

$[b \Rightarrow c \vdash [b \vdash c]](1 := hyp_2, 2 := v)$

▷  { application with $x$ is new }

$[x := hyp_2 \vdash [b \vdash c]](1 := v)$

▷  { removal of definitions}

$[b \vdash c](1 := v)$

▷  { application, removal of definition }

$c$

This demonstrates how reduction "consumes" abstractions without checking for type correctness. The example can be extended still further to yield:

$([u : a \vdash (rules.2(1 := a, 2 := b))(1 := hyp_1, 2 := u)] \therefore [a \vdash b])$ **upto** 1

⬦

$[v : b \vdash (\!\langle\, intro := [p, q : prop \vdash [[p \vdash q] \vdash p \Rightarrow q]]$
$\quad\quad , \; elim \; := [p, q : prop \vdash [p \Rightarrow q \vdash [p \vdash q]]]$
$\quad\quad \rangle\!\rangle.2(1 := b, 2 := c))(1 := hyp_2, 2 := v)$

$]$ **at** 1

▷  { above, structural rules }

$([u : a \vdash (rules.2(1 := a, 2 := b))(1 := hyp_1, 2 := u)] \therefore [a \vdash b])$ **upto** 1

⬦

$[v : b \vdash c]$ **at** 1

▷  { Removal of type tag in judgment }

$[u : a \vdash (rules.2(1 := a, 2 := b))(1 := hyp_1, 2 := u)]$ **upto** 1

⬦

$[v : b \vdash c]$ **at** 1

▷  { cut }

$[\![\, u : a; v := (rules.2(1 := a, 2 := b))(1 := hyp_1, 2 := u)]\!] \vdash c]$

▷  { context linearization }

$[u : a \vdash [v := (rules.2(1 := a, 2 := b))(1 := hyp_1, 2 := u) \vdash c]]$

▷  { removal of definition }

$[a \vdash c]$

Finally, the reduction of a case distinction is illustrated on a text taken from the simplification example (cf. Fig. 12):

$(simplify(1 := b \wedge true))(1 := \langle\!\uparrow((b \wedge true) \wedge true) \mid u\rangle\!)$

▷  { reductions, similar to the ones shown above }

**case** $\langle\!\uparrow((b \wedge true) \wedge true) \mid u\rangle\!$

**of** $\langle\!laws.1(1 := (b \wedge true)), laws.2(1 := (b \wedge true))\rangle\!$

▷  { case distinction }

$(laws.2(1 := (b \wedge true)))(1 := u)$

*Illustration of type assignment:* The first example illustrates type assignment for a subexpression of Fig. 11 (for clarity, the environment is omitted).

$$\mathrm{typ}(rules.2(1 := b, 2 := c))$$
$$= \quad \{ \text{ application } \}$$
$$(\mathrm{typ}(rules.2))(1 := b, 2 := c)$$
$$= \quad \{ \text{ projection } \}$$
$$\mathrm{typ}(rules).2(1 := b, 2 := c)$$
$$= \quad \{ \text{ defined variable } \}$$
$$\langle\; intro := [p, q : prop \vdash [[p \vdash q] \vdash p \Rightarrow q]]$$
$$,\; elim := [p, q : prop \vdash [p \Rightarrow q \vdash [p \vdash q]]]$$
$$\rangle.2(1 := b, 2 := c)$$

It follows from the reductions carried out above that

$$\mathrm{typ}(rules.2(1 := b, 2 := c)) \;\sim\; [b \Rightarrow c \vdash [b \vdash c]]$$

and that

$$\mathrm{typ}((rules.2(1 := b, 2 := c))(1 := hyp_2, 2 := v)) \;\sim\; c$$

Extending the example one step further yields

$$\mathrm{typ}(([u : a \vdash (rules.2(1 := a, 2 := b))(1 := hyp_1, 2 := u)]) \therefore [a \vdash b] \textbf{ upto } 1$$
$$\vartriangleright ([v : b \vdash (rules.2(1 := b, 2 := c))(1 := hyp_2, 2 := v)] \therefore [b \vdash c]) \textbf{ at } 1)$$
$$= \quad \{ \text{ cut } \}$$
$$([u : a \vdash (rules.2(1 := a, 2 := b))(1 := hyp_1, 2 := u)]) \therefore [a \vdash b] \textbf{ upto } 1$$
$$\vartriangleright \mathrm{typ}([v : b \vdash (rules.2(1 := b, 2 := c))(1 := hyp_2, 2 := v)] \therefore [b \vdash c]) \textbf{ at } 1$$
$$= \quad \{ \text{ judgment } \}$$
$$([u : a \vdash (rules.2(1 := a, 2 := b))(1 := hyp_1, 2 := u)]) \therefore [a \vdash b] \textbf{ upto } 1$$
$$\vartriangleright \mathrm{typ}([v : b \vdash (rules.2(1 := b, 2 := c))(1 := hyp_2, 2 := v)]) \textbf{ at } 1$$
$$= \quad \{ \text{ abstraction } \}$$
$$([u : a \vdash (rules.2(1 := a, 2 := b))(1 := hyp_1, 2 := u)] \therefore [a \vdash b]) \textbf{ upto } 1$$
$$\vartriangleright [v : b \vdash \mathrm{typ}((rules.2(1 := b, 2 := c))(1 := hyp_2, 2 := v))] \textbf{ at } 1$$
$$= \quad \{ \text{ applications } \}$$
$$([u : a \vdash (rules.2(1 := a, 2 := b))(1 := hyp_1, 2 := u)] \therefore [a \vdash b]) \textbf{ upto } 1$$
$$\vartriangleright [v : b \vdash (\mathrm{typ}(rules.2)(1 := b, 2 := c))(1 := hyp_2, 2 := v)] \textbf{ at } 1$$
$$= \quad \{ \text{ projection } \}$$
$$([u : a \vdash (rules.2(1 := a, 2 := b))(1 := hyp_1, 2 := u)] \therefore [a \vdash b]) \textbf{ upto } 1$$
$$\vartriangleright [v : b \vdash (\mathrm{typ}(rules).2(1 := b, 2 := c))(1 := hyp_2, 2 := v)] \textbf{ at } 1)$$
$$= \quad \{ \text{ declared variable } \}$$
$$([u : a \vdash (rules.2(1 := a, 2 := b))(1 := hyp_1, 2 := u)] \therefore [a \vdash b]) \textbf{ upto } 1$$
$$\vartriangleright$$

$$[v : b \vdash (\langle\, intro \; := [p, q : prop \vdash [[p \vdash q] \vdash p \Rightarrow q]]$$
$$, \; elim \; := [p, q : prop \vdash [p \Rightarrow q \vdash [p \vdash q]]]$$
$$\rangle.2(1 := b, 2 := c))(1 := hyp_2, 2 := v)$$
$$]\,\mathbf{at}\,1$$

As shown above, the latter expression reduces to $[a \vdash c]$. In other words, the type of the cut in Fig. 11 behaves as required by the judgment.

*Illustration of Auxiliary Predicates.* Based on the illustrations given above it is not difficult to infer that, for instance, the following predicates hold

$$1\text{-dom}(rules.2, prop),$$
$$2\text{-dom}(rules.2, prop),$$
$$1\text{-dom}(rules.2(1 := a, 1 := b), p \Rightarrow q),$$
$$2\text{-dom}(rules.2(1 := a, 1 := b), p).$$

Further, it holds that

$$1\text{-ran}(rules.2(1 := a, 1 := b), [p \vdash q])$$
$$2\text{-ran}(rules.2(1 := a, 1 := b), q)$$

A domain sum occurs in Fig. 10 as follows:

$$\text{dom-sum}(simplify(1 := (b \wedge true), 1 := (b \wedge true)),$$
$$\langle (b \wedge true) \wedge true \mid (b \wedge true) \vee false \rangle).$$

Finally, It is easy to see that

$$\text{pro-size}(rules, 2)$$
$$\text{pro-size}(laws, 2).$$

*Illustration of Explicit Validity.* Both examples of explicit expressions presented in this section are valid. For example, the validity condition for a core judgment of the Syllogism proof (Fig. 11) has been verified above.

## 3.5 The Implicit Part

### 3.5.1 Formation

The implicit part of Deva extends the syntax of the explicit part as follows: The additional grammar rules for texts are

| | | |
|---|---|---|
| $\mathcal{T}$ | $:= \mathcal{T}(\mathcal{T})$ | *(direct application)* |
| | $\mid \; \mathcal{T}(\mathcal{V}_t := \mathcal{T})$ | *(named application)* |
| | $\mid \; \mathcal{T}.\mathcal{V}_t$ | *(named projection)* |
| | $\mid \; \mathcal{T} \; \mathbf{where} \; \mathcal{V}_t := \mathcal{T}$ | *(named product modification)* |
| | $\mid \; \mathcal{T} \diamond \mathcal{T}$ | *(direct cut)* |
| | $\mid \; \mathcal{T} \diamond \mathcal{T} \; \mathbf{at} \; \mathcal{V}_t$ | *(named cut)* |
| | $\mid \; \mathbf{alt}\,[\mathcal{T}, \ldots, \mathcal{T}]$ | *(alternative)* |
| | $\mid \; \mathbf{loop}\,\mathcal{T}$ | *(iteration)* |

and the additional rules for contexts are

$$\mathcal{C} \;:= \mathcal{V}_t?\mathcal{T} \qquad\qquad\qquad (\textit{implicit definition})$$
$$|\; \mathcal{C}(\mathcal{T}) \qquad\qquad\qquad (\textit{direct context application})$$
$$|\; \mathcal{C}(\mathcal{V}_t := \mathcal{T}) \qquad\qquad\qquad (\textit{named context application})$$

### 3.5.2  Intended Meaning of the Constructs

In principle, the valid texts and contexts of the explicit part already constitute a satisfactory means of expression. However, since every construction within an explicitly valid expression must be provided in full detail, using the explicit part as such becomes a too verbose activity for practical use. The operations of the implicit part allow (sometimes quite huge) gaps to be left in developments. Due to the presence of such gaps it becomes impossible to impose type constraints on implicit proofs as done for explicit proofs. However, it is quite possible to impose "implicit" validity constraints, by requiring that implicit operations (i.e., operations with gaps) can be "explained" (Sect. 3.5.9) by explicit operations that satisfy type constraints.

An *implicit definition* $x?t$ declares an identifier $x$ to have the type $t$, just like an ordinary declaration $x : t$ would do. It is called implicit *definition* and not implicit *declaration*, as the reader might have expected, because the explanation process attempts to derive from the context a definition $x := t'$ such that $t'$ is of type $t$, or $t' \therefore t$ for short. The two context applications of the implicit level are variants of the context application of the explicit level. A *direct context application* $c(t)$ instantiates the first declaration in $c$ by the text $t$. A *context application* $c(x := t)$ instantiates the declared or implicitly defined identifier named $x$ in $c$ by $t$. The two text applications of the implicit level , i.e., *direct* and *named application*, are variants of the text operations of the explicit level, analogous to the context applications. Variants of projection and product modification, i.e., *named projection* and *named product modification*, are introduced which allow concrete names to be used for selection, instead of natural numbers. The *direct cut* behaves like the special case of the cut that corresponds to classical functional composition, i.e., $t_1 \diamond t_2$ cuts a function $t_1$ with $t_2$ at the first declaration of $t_2$. The *named cut* $t_1 \diamond t_2$ at $x$ cuts a function $t_1$ with $t_2$ at the declared identifier named $x$.

However, this is not the whole story: All these implicit operations need not satisfy explicit type constraints! Instead, as hinted above, they must be "explainable" by valid explicit operations. Some simple examples may help to illustrate the principle idea behind explaining implicit by explicit expressions. First, assume the environment $\langle x : \mathbf{prim}; y : x; f : [z?\,\mathbf{prim} \vdash [w : z \vdash w]]\rangle$. With respect to this environment, the direct application $f(y)$ is not explicitly valid since the type $x$ of the argument $y$ does not match the parameter type $z$ of the first declaration of the function $f$. However, it is implicitly valid because it can be *explained* by $(f(1 := x))(1 := y)$, which in turn is an explicitly valid text (assuming $f : [z : \mathbf{prim} \vdash [w : z \vdash w]]$). More generally, within a direct application $t_1(t_2)$, implicit definitions on which the first declaration of $t_1$ depends may become instantiated by additional arguments.

For another example, assume the environment $\langle x : \mathbf{prim}; y : [z? \mathbf{prim} \vdash z]; f : [w : \mathbf{prim} \vdash \mathbf{prim}] \rangle$ and consider the direct application $f(y)$. Again, it is not explicitly valid, since the type $[z? \mathbf{prim} \vdash z]$ of $y$ does not conform to the domain $\mathbf{prim}$ of $f$. However, this application is implicitly valid because it can be explained by the explicitly valid text $f(1 := y(1 := x))$. Thus, explanation may also take place by instantiating implicit definitions in the argument $t_2$ of a direct application $t_1(t_2)$.

In addition to omissions of parameters, injections may also be omitted in implicit expressions. Finally, the implicit level introduces two completely new text constructs: The alternative $\mathbf{alt}\,[t_1, \ldots, t_n]$ is a function, which when applied to an argument, selects among the texts $t_1, \ldots, t_n$ for application. The selection must, of course, conform to type of the argument. It must be explained by an appropriate case distinction. The iteration $\mathbf{loop}\,t$ denotes an iterated use of the function $t$ within direct application or direct cuts. It must be explained by an appropriate sequence of direct applications or direct cuts.

### 3.5.3   Environments

Since contexts are extended by three constructs we have to adapt the definition of environments. First, implicit definitions are added to the set of bindings and therefore a binding $X \otimes e$ shall cover also the case of an implicit definition $X?e$. The definition of the lookup function is then changed accordingly and of course $X \oplus x?t = X \oplus \langle x?t \rangle$. We will call a binding $X \otimes e$ an $assumption$ $X \odot e$ if it is either a declaration or an implicit definition. Implicit definitions behave just like declarations with respect to context application:

$$\frac{E \oplus c = E \oplus \langle x?t \rangle \oplus E'}{E \oplus c(1 := t') = E \oplus \langle x := t' \rangle \oplus E'}$$

$$\frac{E \oplus c = E \oplus \langle x?t \rangle \oplus c'}{E \oplus c(n + 1 := t') = E \oplus \langle x?t \rangle \oplus c'(n := t')}$$

A direct context application $c(t)$ instantiates the first declared variable in $c$ to $t$, thus skipping implicit definitions:

$$\frac{E \oplus c = E \oplus \langle x : t \rangle \oplus E'}{E \oplus c(t') = E \oplus \langle x := t' \rangle \oplus E'}$$

$$\frac{E \oplus c = E \oplus \langle x?t \rangle \oplus c'}{E \oplus c(t') = E \oplus \langle x?t \rangle \oplus c'(t')}$$

$$\frac{E \oplus c = E \oplus \langle x := t \rangle \oplus c'}{E \oplus c(t') = E \oplus \langle x := t \rangle \oplus c'(t')}$$

$$\frac{E \oplus c = E \oplus \langle \textbf{context } p := c_1 \rangle \oplus c'}{E \oplus c(t) = E \oplus \langle \textbf{context } p := c_1 \rangle \oplus c'(t)}$$

A named context application $c(x := t)$ instantiates the first assumed variable $x$ in $c$ to $t$:

$$\frac{E \oplus c = E \oplus \langle x \odot t \rangle \oplus E'}{E \oplus c(x := t') = E \oplus \langle x := t' \rangle \oplus E'}$$

$$\frac{E \oplus c = E \oplus \langle y \odot t \rangle \oplus c' \quad x \neq y}{E \oplus c(x := t') = E \oplus \langle y \odot t \rangle \oplus c'(x := t')}$$

$$\frac{E \oplus c = E \oplus \langle y := t \rangle \oplus c'}{E \oplus c(x := t') = E \oplus \langle y := t \rangle \oplus c'(x := t')}$$

$$\frac{E \oplus c = E \oplus \langle \textbf{context } p := c_1 \rangle \oplus c'}{E \oplus c(x := t) = E \oplus \langle \textbf{context } p := c_1 \rangle \oplus c'(x := t)}$$

### 3.5.4 Homomorphic Extensions

Before continuing with the details of the definition, we will introduce a way of defining semantic relations which will save us a considerable amount of space in stating the rules.

While studying the preceding two sections, the reader may have noticed that many of the rules given for the semantic relations could have been easily derived from the structure of the language. Take, for example, the predicate "$\vdash_{val}$" as it is defined in Sect. 3.2.9. The rules given there could be summarized by saying that an expression (i.e., an environment, a text, or a context) is valid if its subexpressions are valid and if some additional relation between the subexpressions holds. It should be sufficient to state only this additional relation for each construct of the language since the rest of the rule is determined by the structure of the construct. This is not quite exact, since we have to distinguish two different types of constructs: *sequential* constructs such as an abstraction, and *parallel* constructs such as an application. These two types induce structurally different rules because subexpressions of sequential constructs may depend on each other whereas subexpressions of parallel constructs do not.

To express this formally, let $\mathcal{SEQ}$ denote the set of sequential phrases specified by the grammar

$$\mathcal{SEQ} := [\mathcal{C}_1 \vdash \mathcal{T}_2] \mid [\mathcal{C}_1; \mathcal{C}_2],$$

let $\mathcal{PAR}_{ex}$ denote the set of parallel phrases of the explicit part specified by the grammar

$$\mathcal{PAR}_{ex} := \mathbf{prim} \mid \mathcal{V}_t \mid \mathcal{T}_1(\mathrm{N} := \mathcal{T}_2) \mid \mathcal{T}_1 \therefore \mathcal{T}_2 \mid$$
$$\langle \mathcal{V}_t := \mathcal{T}_1, \dots, \mathcal{V}_t := \mathcal{T}_k \rangle \mid \mathcal{T}_1.\mathrm{N} \mid \mathcal{T}_1 \textbf{ where } \mathcal{T}_2 := \mathcal{T}_3 \mid$$
$$\langle \mathcal{V}_t := \mathcal{T}_1 \mid \cdots \mid \mathcal{V}_t := \mathcal{T}_k \rangle \mid \mathbf{case } \, \mathcal{T}_1 \mathbf{ of } \mathcal{T}_2 \mid$$
$$\mathcal{T}_1 \textbf{ upto } \mathrm{N} \diamond \mathcal{T}_2 \mathbf{ at } \mathrm{N} \mid$$
$$\mathcal{V}_t : \mathcal{T}_1 \mid \mathcal{V}_t := \mathcal{T}_1 \mid \mathbf{context } \, \mathcal{V}_c := \mathcal{C}_1 \mid \mathbf{import } \, \mathcal{V}_c \mid$$
$$\mathcal{C}_1(\mathrm{N} := \mathcal{T}_2) \mid \mathcal{C}_1[\mathcal{V}_t =: \mathcal{V}_t] \mid \mathcal{C}_1(\mathcal{V}_c =: \mathcal{V}_c],$$

and let $\mathcal{PAR}_{im}$ denote the set of parallel phrases of the implicit part specified by the grammar

$$\mathcal{PAR}_{im} := \mathcal{T}_1(\mathcal{T}_2) \mid \mathcal{T}_1(\mathcal{V}_t := \mathcal{T}_2) \mid$$
$$\mathcal{T}_1.\mathcal{V}_t \mid \mathcal{T}_1 \mathbf{ at } \mathcal{V}_t \mathbf{ to } \mathcal{T}_2 \mid$$
$$\mathcal{T}_1 \diamond \mathcal{T}_1 \mid \mathcal{T}_1 \diamond \mathcal{T}_2 \mathbf{ at } \mathcal{V}_t \mid$$
$$\mathbf{alt }\, [\mathcal{T}_1, \dots, \mathcal{T}_k] \mid \mathbf{loop } \, \mathcal{T}_1 \mid$$
$$\mathcal{V}_t?\mathcal{T}_1 \mid \mathcal{C}_1(\mathcal{T}_2) \mid \mathcal{C}_1(\mathcal{V}_t := \mathcal{T}_2).$$

Here, the indices (i.e., $\mathcal{T}_i$) refer to the argument positions of the subexpressions of the particular construct. Letting $e$ denote arbitrary expressions i.e., either a text or a context, this allows an element of $\mathcal{SEQ}$ to be specified by $c \ominus e$, an element of $\mathcal{PAR}_{ex}$ by $\circledast_{ex}(e_1, \dots, e_k)$ for some $k \geq 0$, and correspondingly an element of $\mathcal{PAR}_{im}$ by $\circledast_{im}(e_1, \dots, e_k)$. Elements of $\mathcal{PAR}_{ex}$ or $\mathcal{PAR}_{im}$ are specified by $\circledast(e_1, \dots, e_k)$ for some $k \geq 0$.

Given a family of predicates $\mathcal{P} = \{\mathcal{P}_\circledast\}$ indexed by the different classes of parallel phrases, where $\mathcal{P}_\circledast(E, e_1, \dots, e_k)$ expresses that $e_1, \dots, e_k$ satisfy relative to the environment $E$ the additional relations associated with the construct $\circledast$. The *homomorphic extension* of $\mathcal{P}$ to environments, texts, and contexts is then specified by the following rules:

$$\frac{}{\vdash_{\mathcal{P}} \langle \rangle} \qquad \frac{E \vdash_{\mathcal{P}} c}{\vdash_{\mathcal{P}} E \oplus c}$$

$$\frac{E \vdash_{\mathcal{P}} c \quad E \oplus c \vdash_{\mathcal{P}} e}{E \vdash_{\mathcal{P}} c \ominus e} \qquad \frac{\vdash_{\mathcal{P}} E \quad E \vdash_{\mathcal{P}} e_1 \cdots E \vdash_{\mathcal{P}} e_k \quad \mathcal{P}_\circledast(E, e_1, \dots, e_k)}{E \vdash_{\mathcal{P}} \circledast(e_1, \dots, e_k)}$$

The first three rules are very obvious with respect to the intuitive meaning of the predicates given above. The last rule is more subtle. Note that it is possible that $k = 0$, for example take $\circledast = \mathbf{prim}$, and hence the condition $\vdash_{\mathcal{P}} E$ is indispensable. Furthermore, the last two rules justify the use of the adjectives "sequential" and "parallel".

As an example, validity $\vdash_{val}$ (as it was defined in Sect. 3.2.9) may be specified as the homomorphic extension of the following family of predicates where we have chosen to write $val_E(\circledast(e_1, \ldots, e_k))$ instead of $val_\circledast(E, e_1, \ldots, e_k)$:

$$
\begin{aligned}
val_E(\mathbf{prim}) &= \text{true}, \\
val_E(x) &= \text{true}, \quad \text{if } E \restriction x \text{ is defined}, \\
val_E(t_1(n := t_2)) &= n\text{-dom}_E(t_1, \text{typ}_E(t_2)), \\
val_E(t_1 \therefore t_2) &= E \vdash \text{typ}_E(t_1) \overset{.}{\sim} t_2, \\
val_E(x : t) &= \text{true}, \\
val_E(x := t) &= \text{true}, \\
val_E(\mathbf{context}\ p := c) &= \text{true}, \\
val_E(\mathbf{import}\ p) &= \text{true}, \quad \text{if } E \restriction p \text{ is defined}.
\end{aligned}
$$

### 3.5.5 Closed Expressions

As a first application, closed expressions that may include constructs of the implicit part are specified as the homomorphic extension of the following family of predicates:

$$
\begin{aligned}
cl_E(x) &= \text{true}, \quad \text{if } E \restriction x \text{ is defined}, \\
cl_E(\mathbf{import}\ p) &= \text{true}, \quad \text{if } E \restriction p \text{ is defined}, \\
cl_E(c[X =: Y]) &= \text{binds}_E(c, X), \\
cl_E(\circledast(e_1, \ldots, e_k)) &= \text{true}, \quad \text{if } \circledast \text{ specifies neither of the above.}
\end{aligned}
$$

### 3.5.6 Extension of Reduction and Explicit Validity

The notion of reduction is extended in a way that is analogous to the way $E \oplus c$ was extended above, i.e., $x : t$ is replaced by $x \odot t$ in all the axioms defining the elementary reduction steps. The structural rules and the closure rules are not changed. Similarly, the auxiliary semantic predicates from the explicit part are extended by replacing $x : t$ everywhere by $x \odot t$. This yields an extension of explicit validity that includes implicit definitions. Note that those are treated in *exactly* the same way as declarations.

### 3.5.7 Auxiliary Semantic Predicates for Implicit Validity

The implicit part introduces a number of constructs that are slight variants of already existing explicit constructs, i.e., direct application, named application, and so on. In this section, a number of auxiliary relations are specified and then used to relate these constructs of the implicit part to constructs of the explicit part.

From the explicit part, one already knows the auxiliary relation *n-th domain*. Several variations of this relation are introduced in the implicit part. The first such variation is the relation *declaration position*. Figuratively speaking, a text $t$ has declaration position $n$ relative to an environment $E$, denoted by dec-pos$_E(t, n)$, if the $n$-th assumption in $t$ is a declaration, i.e., it is not an implicit definition. This is made precise by the following rules:

$$\frac{E \vdash \; \text{typ}_E(t) \overset{\sim}{\;} [x : t_1 \vdash t_2]}{\text{dec-pos}_E(t, 1)} \qquad \frac{E \vdash \; t \overset{\sim}{\;} [x : t_1 \vdash t_2]}{\text{dec-pos}_E(t, 1)}$$

$$\frac{E \vdash \; \text{typ}_E(t) \overset{\sim}{\;} [x \odot t_1 \vdash t_2] \quad \text{dec-pos}_{E \oplus x \odot t_1}(t(1 := x), n)}{\text{dec-pos}_E(t, n + 1)}$$

$$\frac{E \vdash \; t \overset{\sim}{\;} [x \odot t_1 \vdash t_2] \quad \text{dec-pos}_{E \oplus x \odot t_1}(t_2, n)}{\text{dec-pos}_E(t, n + 1)}$$

Note, that the structure of the rules is the same as that of the rules defining the predicate $n$-dom$_E(t_1, t_2)$.

The *minimal declaration position*, denoted by min-dec-pos$_E(t, n)$, is defined as a predicate that holds if and only if $n$ is the smallest $m$ such that the predicate dec-pos$_E(t, m)$ is true. As will be seen later, this relation allows to precisely relate e.g. the direct application of the implicit part to the application of the explicit part.

Also, there exist variants of the above relations for contexts, denoted by dec-pos$_E(c, n)$ and min-dec-pos$_E(c, n)$. By replacing the two declarations in the above rules by implicit definitions one obtains the rules of the relation *implicit definition position*, denoted by idef-pos$_E(t, n)$, and consequently all the other relations, i.e., min-idef-pos$_E(t, n)$ and the corresponding relations on contexts. These formal definitions are quite obvious and are therefore omitted.

The named applications and named cuts of the implicit part which specify the parameter to be instantiated by giving its name, e.g. $t_1(x := t_2)$, are related to the application and cut of the explicit part by the predicate *assumption position*. Figuratively speaking, a text $t$ has an $x$-assumption position $n$ relative to an environment $E$, denoted by $x$-asm-pos$_E(t, n)$, if the $n$-th binding in $t$ is an assumption $x \odot t'$. Unfortunately, it turns out that there exist cases where the identifier used in the $n$-th assumption differs, depending on whether $t$ or typ$_E(t)$ is considered. For example, take the text

$$[z : [x : \textbf{prim} \vdash \textbf{prim}] \vdash z](1 := [y : \textbf{prim} \vdash y])$$

which indeed has either $x : \textbf{prim}$ or $y : \textbf{prim}$ as 1-st assumption position, depending on whether its type is considered or not. Such problematic cases are excluded by the predicate asm-unique$_E(t)$ which is true if and only if $E \vdash \; t \sim [x \odot t_1 \vdash t_2]$ and $E \vdash \; \text{typ}_E(t) \sim [y \odot t_1' \vdash t_2']$ implies that $x = y$. This predicate can be algorithmically realized by simply inspecting the first assumptions in the "normal forms" of $t$ and typ$_E(t)$. Now, the predicate $x$-assumption position can be specified by the following rules:

$$\frac{E \vdash \text{typ}_E(t) \sim [x \odot t_1 \vdash t_2] \quad \text{asm-unique}_E(t)}{x\text{-asm-pos}_E(t, 1)}$$

$$\frac{E \vdash t \sim [x \odot t_1 \vdash t_2] \quad \text{asm-unique}_E(t)}{x\text{-asm-pos}_E(t, 1)}$$

$$\frac{E \vdash \text{typ}_E(t) \sim [y \odot t_1 \vdash t_2] \quad x\text{-asm-pos}_{E \oplus y \odot t_1}(t(1 := y), n)}{x\text{-asm-pos}_E(t, n + 1)}$$

$$\frac{E \vdash t \sim [y \odot t_1 \vdash t_2] \quad x\text{-asm-pos}_{E \oplus y \odot t_1}(t_2, n)}{x\text{-asm-pos}_E(t, n + 1)}$$

Once again, the *minimal assumption position*, denoted by $x\text{-min-asm-pos}_E(t, n)$, is defined as a predicate that holds if and only if $n$ is the smallest $m$ such that $x\text{-asm-pos}_E(t, m)$ holds. Again, there exists variants of this predicate for contexts (these are denoted by $x\text{-asm-pos}_E(c, n)$ and $x\text{-min-asm-pos}_E(c, n)$). Their definition is obvious and therefore omitted.

The product operations of the implicit part which specify the component of interest by giving its name, e.g. $t.x$, are related to the product operations of the explicit part by the predicate *component position*. Figuratively speaking, a text $t$ has *an* $x$-component position $n$ relative to an environment $E$, denoted by $x\text{-com-pos}_E(t, n)$, if $t$ yields a product $[x_1 := t_1, \ldots, x := t_n, \ldots, x_m := t_m]$. Note that similar to assumption names it is well possible that component names can change, depending on whether a text or its type is considered. For example take the text,

$$[z : \langle x := \mathbf{prim}, y := \mathbf{prim} \rangle \vdash z](1 := \langle y := w, x := w \rangle)$$

(where $w : \mathbf{prim}$) which has either $x$ or $y$ as its first component name, depending on whether its type is considered or not. Again there is a predicate $\text{com-unique}_E(t)$ to enforce the uniqueness of the component names of $t$ and $\text{typ}_E(t)$. It is true if and only if $E \vdash t \sim [\ldots, x := t, \ldots]$ and $E \vdash \text{typ}_E(t) \sim [\ldots, y := t', \ldots]$ implies that $x = y$. The predicate $x$-component position is now made precise by the following rules:

$$\frac{E \vdash \text{typ}_E(t) \sim [x_1 := t_1, \ldots, x := t_n, \ldots, x_m := t_m] \quad \text{com-unique}_E(\text{typ}_E(t))}{x\text{-com-pos}_E(t, n)}$$

$$\frac{E \vdash t \sim [x_1 := t_1, \ldots, x := t_n, \ldots, x_m := t_m] \quad \text{com-unique}_E(t)}{x\text{-com-pos}_E(t, n)}$$

The *minimal component position*, denoted by $x\text{-min-com-pos}_E(t, n)$, is defined as a predicate that holds if and only if $n$ is the smallest $m$ such that the predicate $x\text{-com-pos}_E(t, m)$ is true.

### 3.5.8  Explicitation

A central semantic relation of the implicit part describes the *explicitation* ,
of an expression $e_1$ by an expression $e_2$ under an environment $E$, denoted by
$E \vdash e_1 \sqsupseteq e_2$. Intuitively, $e_1$ and $e_2$ both are explicitly valid except for one
operation, that may be either implicit or explicit, and $e_2$ is "closer" to explicit
validity than $e_1$.

The laws specifying explicitation come in several groups. The first group of
explicitation rules describes the direct assignment of an explicit construct to an
implicit construct. The first four rules deal with applications and projections.
These rules make use of the auxiliary predicates just introduced (Sect. 3.5.7) to
formalize the intended meaning (Sect. 3.5.2) of those implicit constructs that are
close variants of existing constructs of the explicit part.

(E1)
$$\frac{\text{min-dec-pos}_E(e, n)}{E \vdash \ e(t) \sqsupseteq e(n := t)}$$

(E2)
$$\frac{x\text{-min-asm-pos}_E(e, n)}{E \vdash \ e(x := t) \sqsupseteq e(n := t)}$$

(E3)
$$\frac{x\text{-min-com-pos}_E(t, n)}{E \vdash \ t.x \sqsupseteq t.n}$$

(E4)
$$\frac{x\text{-min-com-pos}_E(t_1, n)}{E \vdash \ t_1 \textbf{ where } x := t_2 \sqsupseteq t_1 \textbf{ where } n := t_2}$$

*Remark.* Explanation will be defined by applying explicitation recursively on
the structure of fully implicit expressions. Therefore, in these rules it can be
assumed that the arguments of $\circledast_{im}$ have been already explained, i.e., they are
explicitly valid. Therefore, it makes sense to use the auxiliary predicates.

The rules for cuts are quite similar, except that the **upto** parameter is
deliberately chosen to be any natural $m$ for which an $m$-th range exists in the
first argument.

(E5)
$$\frac{\exists t : m\text{-ran}_E(t_1, t) \quad \text{min-dec-pos}_E(t_2, n)}{E \vdash \ t_1 \lozenge t_2 \sqsupseteq t_1 \textbf{ upto } m \lozenge t_2 \textbf{ at } n}$$

(E6)
$$\frac{\exists t : m\text{-ran}_E(t_1, t) \quad x\text{-min-asm-pos}_E(t_2, n)}{E \vdash \ t_1 \lozenge t_2 \textbf{ at } x \sqsupseteq t_1 \textbf{ upto } m \lozenge t_2 \textbf{ at } n}$$

The last rule shows how an alternative can be explained by a case distinction,
in case *all* its components start with a declared first parameter.

(E7)
$$\frac{\text{dec-pos}_E(t_i, 1) \quad 1\text{-val-dom}_E(t_i, t_i')}{E \vdash \ \textbf{alt} \, [t_1, \dots, t_n]}$$
$$\sqsupseteq [x : \langle y := t_1' \mid \dots y := t_n' \rangle \vdash \textbf{case } x \textbf{ of } \langle y := t_1, \dots, y := t_n \rangle]$$

The notation $n$-val-dom$_E(e, t)$ is shorthand for $n$-dom$_E(e, t), E \vdash_{val} t$. Note that these rules cannot guarantee the validity constraints of the explicit construct by which the implicit construct has been explained. In fact, there still remain gaps in the explicit operation and therefore these constraints are usually not satisfied. The purpose of the next group of rules is to describe in what way these gaps can be "filled". The first rule describes the extraction of implicitly defined parameters from the arguments of an implicit construct.

$$
\text{(E8)} \quad
\frac{\text{idef-pos}_E(e_i, n) \quad x\text{-asm-pos}_E(e_i, n) \\ n\text{-val-dom}_E(e_i, t) \quad x\text{-free}(i, \circledast_{im}(e_1, \ldots, e_m))}{\begin{array}{l} E \vdash \circledast_{im}(e_1, \ldots, e_i, \ldots, e_m) \\ \sqsupseteq (x?t \ominus \circledast_{im}(e_1, \ldots, e_i(n := x), \ldots, e_m)) \end{array}}
$$

*Remark.* Note that the extracted implicit definition $x?t$ is logically "connected" with $e_i$ via the application $e_i(n := x)$ whose explicit validity follows easily from the preconditions of the rule.

The condition $x\text{-free}(i, \circledast_{im}(e_1, \ldots, e_n))$ serves to prohibit certain paradoxical cases in which the collection of an implicit definition of an identifier $x$ from a particular argument of $\circledast_{im}$ would result in a clash with the meaning of $\circledast_{im}$. For example, the named application $[x?t_1 \vdash t_2](x := t_3)$ should be explicitated into the application $[x?t_1 \vdash t_2](1 := t_3)$ only and not into the abstraction $[x?t_1 \vdash (t_2(1 := x))(x := t_3)]$. The condition is defined as follows:

$$
\begin{aligned}
x\text{-free}(1, t_1(y := t_2)) &= (x \neq y) \\
x\text{-free}(2, t_1 \oslash t_2 \text{ at } y) &= (x \neq y) \\
x\text{-free}(1, c(y := t)) &= (x \neq y) \\
x\text{-free}(i, \circledast_{im}(e_1, \ldots, e_n)) &= true \text{ otherwise}
\end{aligned}
$$

In order to allow multiple extractions of implicit definitions, the following monotonicity rule for explicitation is added.

$$
\text{(E9)} \quad
\frac{E \oplus \langle x?t \rangle \vdash e_1 \sqsupseteq e_2}{E \vdash (x?t \ominus e_1) \sqsupseteq (x?t \ominus e_2)}
$$

Another extraction rule allows to extract declared parameters within the arguments of an explicit construct.

$$
\text{(E10)} \quad
\frac{\text{dec-pos}_E(e_i, 1) \quad x\text{-asm-pos}_E(e_i, 1) \quad 1\text{-val-dom}_E(e_i, t)}{\begin{array}{l} E \vdash \circledast_{ex}(e_1, \ldots, e_i, \ldots, e_n) \\ \sqsupseteq \circledast_{ex}(e_1, \ldots, [x : t \vdash e_i(1 := x)], \ldots, e_n) \end{array}}
$$

Note that this rule is reminiscent of the reverse the "eta contraction" principle, i.e.,

$$
[x : t_1 \vdash t_2(1 := x)] \rhd t_2,
$$

provided $x$ does not occur free in $t_2$.

Again, in order to allow multiple extractions, the following monotonicity rule for explicitation is added:

$$\frac{E \oplus \langle x : t \rangle \vdash \circledast_{ex}(e'_1, \dots, e'_i, \dots, e'_n) \sqsupseteq \circledast_{ex}(e_1, \dots, e_i, \dots, e_n) \qquad \forall j : j \neq i \bullet E \vdash_{cl} e'_j}{E \vdash \circledast_{ex}(e'_1, \dots, [x : t \vdash e'_i], \dots, e'_n) \sqsupseteq \circledast_{ex}(e_1, \dots, [x : t \vdash e_i], \dots, e_n)}$$

(E11)

The following rule specifies that explicitation right-commutes with conversion restricted to explicitly valid expressions.

(E12)
$$\frac{E \vdash e_1 \sqsupseteq e_2 \quad E \vdash e_2 \sim e_3 \quad E \vdash_{val} e_2 \quad E \vdash_{val} e_3}{E \vdash e_1 \sqsupseteq e_3}$$

The next rule is the *essential* one, because it describes how an extracted implicit definition may be replaced by an appropriate explicit definition.

(E13)
$$\frac{E \vdash_{val} t_2 \quad E \vdash \text{typ}_E(t_2) \sim t_1}{E \vdash (x?t_1 \ominus e) \sqsupseteq (x := t_2 \ominus e)}$$

Note the nondeterminism of this rule, i.e., *any* valid text $t_2$ of type $t_1$ is allowed for the instantiation.

Finally, there is a rule that describes the injection of an expanded argument into a sum with arbitrary (explicitly valid) components.

(E14)
$$\frac{\forall j : 1 \leq j \leq k, j \neq i \bullet E \vdash_{val} t_j}{\begin{array}{l} E \vdash \circledast_{ex}(e_1, \dots, e_i, \dots, e_n) \\ \qquad \sqsupseteq \circledast_{ex}(e_1, \dots, \langle x_1 := {}^{\frown r} t_1 \mid \cdots \mid x_i := e_i \mid \cdots \mid x_k := {}^{\frown r} t_k \rangle, \dots, e_n) \end{array}}$$

Together with obvious rules for reflexivity and transitivity, this concludes the definition of explicitation.

### 3.5.9   Explanation of Expressions

In this section, explicitation is lifted to *explanation*, denoted by $E \vdash e_1 \rightsquigarrow e_2$. Intuitively, to explain an implicit operation amounts to first explaining its subexpressions and then explicitating the operation on the explained subexpressions into a valid explicit expression. This is made precise by the following rule:

$$\frac{\vdash_{val} E \quad E \vdash e_1 \rightsquigarrow e'_1 \quad \cdots \quad E \vdash e_n \rightsquigarrow e'_n \qquad E \vdash \circledast_{im}(e'_1, \dots, e'_n) \sqsupseteq e \quad E \vdash_{val} e}{E \vdash \circledast_{im}(e_1, \dots, e_n) \rightsquigarrow e}$$

Note how the validity constraint serves as a *filter*, i.e., only explicitations into explicitly valid expressions are allowed. This process is the implicit analogue to the validity constraints of the explicit part.

The above rule only deals with implicit operations. The following rule states that explicit constructs must not be explained by something other than themselves, however their validity constraint must be met.

$$\frac{\vdash_{val} E \quad E \vdash e_1 \rightsquigarrow e_1' \quad \cdots \quad E \vdash e_n \rightsquigarrow e_n' \quad val_E(\circledast_{ex}(e_1', \ldots, e_n'))}{E \vdash \circledast_{ex}(e_1, \ldots, e_n) \rightsquigarrow \circledast_{ex}(e_1', \ldots, e_n')}$$

Note that for explicit operations without subexpressions (i.e., $n = 0$) the above schematic inference rule becomes an axiom. Sequential operations do not have validity constraints and hence they must not be explained. Therefore they allow the following monotonicity rule for explanation.

$$\frac{E \vdash c \rightsquigarrow c' \quad E \oplus c' \vdash e \rightsquigarrow e'}{E \vdash c \ominus e \rightsquigarrow c' \ominus e'}$$

The definition of explanation is still incomplete, because iterations are not yet handled. Iterations cannot be explained as such, but only when occurring in a situation together with certain other constructs. One such situation is $(\mathbf{loop}\, t_1)(t_2)$, i.e., the text $t_1$ is applied repeatedly, possibly zero times, to $t_2$. The explanation of this situation is not very surprising:

$$\frac{E \vdash t_1(t_1(\ldots t_1(t_2)\ldots)) \rightsquigarrow t_3}{E \vdash (\mathbf{loop}\, t_1)(t_2) \rightsquigarrow t_3}$$

Another situation is $t_1 \lozenge \mathbf{loop}\, t_2$, i.e., the text $t_2$ is cut repeatedly with $t_1$. The explanation of this situation is as follows:

$$\frac{E \vdash (\cdots((t_1 \lozenge t_2) \lozenge t_2) \cdots \lozenge t_2) \rightsquigarrow t_3}{E \vdash t_1 \lozenge \mathbf{loop}\, t_2 \rightsquigarrow t_3}$$

These two rules complete the definition of explanation. It is easy to check that all explanations are explicitly valid.

*Remark.* Explicitation and explanation can be seen as constructive but highly nondeterministic operations. Thus, they are not very practical from the point of view of an implementation. A much more practical approach is to guide explicitation by constraints extracted from the expression to be explicitated. Such a guided explicitation process is defined and studied for a kernel subset of Deva in [101].

## 3.5.10  Implicit Validity

The notion of implicit validity relative to a valid environment $E$, denoted by $E \vdash_{imval} e$, is defined as the range of explanation relative to $E$.

$E \vdash_{imval} e$ if and only if $\vdash_{val} E$ and there exists an $e'$ with $E \vdash e \rightsquigarrow e'$.

## 3.6   The Implicit Part: Illustrations

In the following illustrations we refer to Fig. 9 and Fig. 10. In the context of
Fig. 9 it is easy to infer that the predicates

$$\text{dec-pos}(rules.2, 3),$$
$$\text{dec-pos}(rules.2, 4),$$
$$\text{min-dec-pos}(rules.2, 3),$$

are true. In the context of Fig. 9, the predicates

$$p\text{-asm-pos}(rules.2, 1),$$
$$q\text{-asm-pos}(rules.2, 2),$$
$$p\text{-min-asm-pos}(rules.2, 1),$$
$$and\text{-com-pos}(laws, 1),$$
$$or\text{-com-pos}(laws, 2).$$

hold.

A very simple kind of explicitation is the replacement of names by numbers
in named operations, e.g. assuming the context from Fig. 9:

$rules.elim$

⊒   { E3, since $elim\text{-com-pos}(rules, 2)$ }

$rules.2$

Frequently, such a direct mapping is insufficient to ensure explicit validity. Typically, additional applications need to be introduced as in the following explicitation.

$rules.2(hyp_2)$

⊒   { E8, where $e_1 := rules.2$, $e_2 := hyp_2$, $n := 1$, and $t := prop$ }

$[p?prop \vdash (rules.2(1 := p))(hyp_2)]$

⊒   { E13, where $t_1 := prop$ and $t_2 := b$ }

$[p := b \vdash (rules.2(1 := p))(hyp_2)]$

⊒   { E12, based on an obvious conversion }

$(rules.2(1 := b))(hyp_2)$

⊒   { E8, where $e_1 := rules.2(1 := b)$, $e_2 := hyp_2$, $n := 1$, and $t := prop$ }

$[q?prop \vdash (rules.2(1 := b, 2 := q))(hyp_2)]$

⊒   { E13, where $t_1 := prop$ and $t_2 := c$ }

$[q := c \vdash (rules.2(1 := b, 2 := q))(hyp_2)]$

⊒   { E12, based on an obvious conversion }

$(rules.2(1 := b, 2 := c))(hyp_2)$

⊒   { E1, since $\text{min-dec-pos}(rules.2(1 := b, 2 := c), 1)$ }

$(rules.2(1 := b, 2 := c))(1 := hyp_2)$

To clarify understanding, some instantiations of rule parameters have been given. Note that choosing $b$ and $c$ as arguments of the two new applications is the only possibility to ensure explicit validity. The explicitation of the cut in the Syllogism proof can be illustrated as follows:

$(rules.2(1 := a, 1 := c))(1 := hyp_1)$
$\diamond(rules.2(1 := b, 1 := c))(1 := hyp_2)$
$\sqsupseteq$  { E5 }
$(rules.2(1 := a, 1 := c))(1 := hyp_1)$ **upto** 1
$\diamond(rules.2(1 := b, 1 := c))(1 := hyp_2)$ **at** 1
$\sqsupseteq$  { E10 twice }
$[u : a \vdash (rules.2(1 := a, 1 := c))(1 := hyp_1, 1 := u)]$ **upto** 1
$\diamond[v : b \vdash (rules.2(1 := b, 1 := c))(1 := hyp_2, 1 := v)]$ **at** 1

Note that explicitation is always concerned with a single operation only. The arguments of that operation must be explicitly valid expressions. The explicitation of the alternative in Fig. 10 runs as follows:

**alt**$\langle laws.1, laws.2 \rangle$
$\sqsupseteq$  { E8 }
$[a? bool \vdash$ **alt**$\langle laws.1(1 := a), laws.2 \rangle$
$\sqsupseteq$  { E8, in combination with structural rule E9 }
$[a? bool \vdash [b? bool \vdash$ **alt**$\langle laws.1(1 := a), laws.2(1 := b) \rangle]]$
$\sqsupseteq$  { E13, in combination with E9 }
$[a? bool \vdash [b := a \vdash$ **alt**$\langle laws.1(1 := a), laws.2(1 := b) \rangle]]$
$\sqsupseteq$  { E12 }
$[a? bool \vdash$ **alt**$\langle laws.1(1 := a), laws.2(1 := a) \rangle]]$
$\sqsupseteq$  { E7, in combination with structural rule E9 }
$[$  $a? bool; v : \langle a \wedge true \mid a \vee false \rangle$
$\vdash$ **case** $v$ **of** $\langle laws.1(1 := a), laws.2(1 := a) \rangle$
$]$

Using the result of this explicitation (corresponding to the definition of *simplify* in Fig. 10), one can illustrate the explicitation of a single simplification step.

$simplify(u)$
$\sqsupseteq$  { E8 }
$[a? bool \vdash simplify(1 := a)](u)$
$\sqsupseteq$  { E13 }
$[a := (b \wedge true) \vdash simplify(1 := a)](u)$
$\sqsupseteq$  { E12 }
$(simplify(1 := (b \wedge true)))(u)$

$\sqsupseteq$  { E1 }

$(simplify(1 := (b \wedge true)))(1 := u)$

$\sqsupseteq$  { E14, where $e_1 := simplify(1 := (b \wedge true))$, $e_2 := u$, and $r := 1$ }

$(simplify(1 := (b \wedge true)))(1 := \langle\hat{}((b \wedge true) \wedge true) \mid u\rangle)$

The reduction of the latter expression has been illustrated in Sect. 3.4.

Fig. 9 and Fig. 10 are explanations of Fig. 11 and Fig. 12 respectively. Some of the explicitation steps of these explanations have been illustrated above.

## 3.7   Mathematical Properties of Deva

This section summarizes basic parts of the language theory of Deva as defined in the previous section. The detailed proofs can be found in [104]. There are essentially three "desirable" properties of the language theory:

- Reduction should be *confluent*, i.e., whenever an expression reduces to two different expressions, then there should be an expression to which these two expressions can be reduced. As a consequence, the result of reduction is independent of the chosen strategy for applying the reduction axioms, e.g. lazy-evaluation, and thus implementations of reduction using any such strategy are guaranteed to be partially correct.
- Explicit validity should be *closed* against reduction and typing, i.e., whenever something is reduced or typed, it should not cease being explicitly valid. This property is quite important for procedures that check explicit validity, it assures that explicit validity must not be rechecked after applying a reduction rule or after computing a type.
- Finally, reduction should have the *strong normalization property* on explicitly valid expressions, i.e., all reduction chains starting from an explicitly valid expressions should terminate. Quite obviously, this property is essential for the termination of validity checking procedures.

It will turn out that the first two properties can be proven for Deva in full. However, strong normalization can only be proven under a slight restriction (see Sect. 3.7.3) and it remains an open problem whether it is satisfied in full.

These three desirable properties are fundamental for many further results. As an illustration of the utilization of these properties two important practical consequences will be briefly presented:

- A decision procedure for explicit validity is outlined.
- It is shown how the problem to prove the adequacy of theory formalizations in Deva can be tackled.

### 3.7.1   Confluence

The first desirable property, i.e., confluence of reduction, can be stated more precisely as follows:

**Theorem 1 (Confluence of reduction).** *Consider an arbitrary environment E and an expression e which is closed under E: If there are expression $e_1$ and $e_2$ such that $E \vdash e \triangleright e_1$ and $E \vdash e \triangleright e_2$, then $e_1$ and $e_2$ have a common reduct, i.e., there exists an expression $e'$ such that $E \vdash e_1 \triangleright e$ and $E \vdash e_2 \triangleright e$.*

**Proof (Outline).** The basic idea to prove confluence is to partition reduction into a number of subreductions, prove confluence of each of these subreductions, and then prove that their combination preserves the property. The following partition into subreductions is considered:

**Definition 2 (Partition of $\triangleright$).** Reduction can be partitioned into the following three subreductions:

   a. Let $\triangleright_\delta$ denote the restriction of $\triangleright$ to the axiom for unfolding text and context definitions (see 3.2.5).
   b. Let $\triangleright_s$ denote the restriction of $\triangleright$ to the axioms to split context joins as left arguments of abstractions and context joins.
   c. Let $\triangleright_r$ denote the restriction of $\triangleright$ to all axioms except those of of $\triangleright_\delta$ and $\triangleright_s$.

By a technical, but straightforward analysis, each of these three reductions can be shown to be confluent, i.e., all of them are locally confluent and have the strong normalization property. Furthermore, it can be shown that these reductions *commute* with each other. For example, if $E \vdash e_1 \triangleright_\delta e_2$ and $E \vdash e_1 \triangleright_s e_3$ then $E \vdash e_2 \triangleright_s e_4$ and $E \vdash e_3 \triangleright_\delta e_4$. The confluence of reduction follows from the simple fact that confluence is preserved by the union of reductions that commute.  □

It is interesting to note that in contrast to the well-known confluence proof for $\lambda$-calculus described e.g. in [52], the proof in [104] which was outlined above does not introduce the notion of *parallel reduction*. The necessity to introduce such a reduction arises from the *simultaneous* substitution of all free occurrences of the formal parameter $x$ by the argument $t$ in the $\beta$-conversion principle

$$(\lambda x.t)(t') \sim t'[x/t]$$

of $\lambda$-calculus. In contrast, the internalized definition mechanism of Deva allows for the *individual* substitution of free occurrences by $\triangleright_\delta$ steps. This substitution process is more fine-grain than that in $\lambda$-calculus. As a consequence, confluence can be shown by straightforward decomposition as hinted above.

An easy and well-known corollary to confluence is the Church-Rosser property.

**Corollary 3 (Reduction is Church-Rosser).** *Consider an arbitrary environment E and expressions $e_1$, $e_2$ which are closed under E: If $E \vdash e_1 \sim e_2$ then there exists a common reduct of $e_1$ and $e_2$.*

### 3.7.2  Closure Results

The second desirable property, i.e., the closure results, can be stated more precisely as follows:

**Theorem 4 (Closure against typing and reduction).** *Consider an arbitrary environment $E$ and expressions $e_1$ and $t$ which are explicitly valid under $E$, i.e., $E \vdash_{val} e_1$ and $E \vdash_{val} t$. The two following properties hold:*

- *If $E \vdash e_1 \triangleright e_2$ then $E \vdash_{val} e_2$, and*
- *if $typ_E(t)$ is defined then $E \vdash_{val} typ_E(t)$.*

**Proof (Outline).** Closure against reduction is shown by proving that all reduction axioms preserve closure and that the structural rules of reduction preserve closure. These very technical investigations are based on several substitution lemmas for reduction, typing, and validity.

Closure against typing essentially has to check whether the validity restrictions of a construct are preserved by typing the construct. It turns out that this can be shown using an additional lemma, if $E \vdash t_1 \sim t_2$ then $E \vdash typ_E(t_1) \sim typ_E(t_2)$ where $t_1$ and $t_2$ must be valid. This lemma follows from the first closure result. □

### 3.7.3 Strong Normalization

It is easy to see that reduction does not necessarily terminate on arbitrary closed expressions, for example the expression

$$[d := [x : \mathbf{prim} \vdash x(1 := x)] \vdash d(1 := d)]$$

produces a nonterminating chain of reduction steps. However, the type restrictions imposed upon explicitly valid expressions are designed to exclude such expressions. For example, in the above expression the application $x(1 := x)$ is not explicitly valid because $val_E(x(1 := x))$ implies the existence of a "domain" of $x$ (which is clearly non-existent). In fact, explicitly valid expressions should be *strongly normalizable*:

**Definition 5 (Strongly normalizable).** An expression $e$ is strongly normalizable under E if and only if there is no infinite chain of reductions of the following shape:

$$E \vdash e \triangleright^1 e_1 \triangleright^1 e_2 \triangleright^1 \ldots \triangleright^1 e_i \triangleright^1 \ldots$$

It turns out that the essential condition that is violated by the above example is the condition that a first domain exists, regardless whether it conforms to the type of the argument or not. In fact, strong normalization can be shown to hold for a much wider class of expressions, the *normable* expressions, in which certain generalized structural conditions have to be satisfied. It can be shown that explicitly valid expressions are normable, except for expressions which contain case distinctions. Case distinctions pose a problem because it turns out to be extremely complex to *statically*, i.e., without performing reduction, figure out which case is eventually selected, in a case distinction in a given expression. However, since the overall structure of the expression depends on this selection, the concept of normability could not be defined for expressions containing case distinctions. In the following discussions, the adjective *case-free* will be used to describe expressions and environments that do not contain case distinctions.

The next step of the strong normalization consists of the proof that any normable expression is indeed strongly normalizable. This can be shown with an adaptation of a computability proof presented in [100] for a related system. The idea is to restrict the set of expressions from which reduction terminates, yielding the set of expressions *computable under substitution*, in such way that it can be shown by simple structural induction that all normable expressions are computable under substitution and therefore strongly normalizable. By the previous discussion, this implies the strong normalization of case-free explicitly valid expressions.

**Theorem 6 (Strong normalization of case-free explicit validity).** *For any expression e valid under an environment E, i.e., $E \vdash_{val} e$, it holds that if e and E are case-free then e is strongly normalizable under E.*

### 3.7.4 Decidability of Case-Free Validity

Because of the confluence and the strong normalization results, the conversion relation between case-free valid expressions can be decided in a very simple way. To illustrate this point, the following theorem is needed:

**Theorem 7 (Unique normal form).** *For any expression e valid under an environment E, i.e., $E \vdash_{val} e$, it holds that if e and E are case-free then any reduction of e terminates and moreover leads to the same expression. This unique result of reduction is denoted by $NF_E(e)$ (normal form).*

**Proof.** By strong normalization, the reduction of $e$ indeed terminates and by confluence it produces a unique result. $\square$

To check whether $E \vdash e_1 \sim e_2$ holds for case-free expressions $e_1$, $e_2$ that are valid under the case-free environment $E$, it suffices to check whether $NF_E(e_1) = NF_E(e_2)$. This is correct since by Church-Rosser there exists an expression $e$ such that both $e_1$ and $e_2$ reduce to $e$. By closure, $e$ is valid (and obviously case-free). Thus, $e_1$, $e_2$, and $e$ each have unique normal forms which must be obviously identical. This procedure can be extended in a simple (but technical) way to a decision procedure for case-free validity:

**Theorem 8 (Decidability of case-free explicit validity).** *For any case-free environment E and expression e it is decidable whether $E \vdash_{val} e$.*

The above procedure to decide conversion between two expressions requires to compute the normal forms of the expressions. However, the size of the normal form of an expression may explode exponentially with respect to the size of the original expression, e.g. by complete unfolding of definitions. Thus the procedure is not very practical. There are many ways to develop more efficient procedures to decide conversion, based on the fact that to prove $e_1 \sim e_2$ it suffices to find a (possibly still reducible) common reduct of $e_1$ and $e_2$. However, the correctness of such a procedure always depends on the properties of confluence and closure (for partial correctness), and strong normalization (for termination).

### 3.7.5 Recursive Characterization of Valid Normal Forms

Given the existence and uniqueness of normal forms for explicitly valid expressions, it is interesting to look at the structure of these normal forms. The following definition proposes a recursive characterization of valid normal forms.

**Definition 9 (Valid normal forms).** The set of valid normal forms, denoted by $\mathcal{VNF}_E$, and the set of "dead ends" i.e., the set of all valid expressions with irreducible elimination constructs at the top, denoted by $\mathcal{DE}_E$, are defined mutually recursively as follows:

$$
\begin{aligned}
\mathcal{VNF}_E \;:=\; &\{X \odot e & &| \; e \in \mathcal{VNF}_E\} \\
&\cup \;\{[\![\, X \odot e; c\,]\!] & &| \; e \in \mathcal{VNF}_E, c \in \mathcal{VNF}_{E \oplus \langle X \odot e \rangle}\} \\
&\cup \;\{\mathbf{prim}\} \\
&\cup \;\{[x \odot t_1 \vdash t_2] & &| \; t_1 \in \mathcal{VNF}_E, t_2 \in \mathcal{VNF}_{E \oplus \langle x \odot t_1 \rangle}\} \\
&\cup \;\{\langle\!| x_1 := t_1, \ldots, x_k := t_k |\!\rangle & &| \; t_i \in \mathcal{VNF}_E \text{ for } i = 1, \ldots, n\} \\
&\cup \;\{\langle\!| x_1 := \overset{r}{\phantom{.}}\!{}^{\frown} t_1 \mid \cdots \mid x_i := t_i \mid \cdots \mid x_k := \overset{r}{\phantom{.}}\!{}^{\frown} t_k |\!\rangle \\
& & &| \; t_i \in \mathcal{VNF}_E \text{ for } i = 1, \ldots, n\} \\
&\cup \;\mathcal{DE}_E
\end{aligned}
$$

$$
\begin{aligned}
\mathcal{DE}_E \;:=\; &\{x & &| \; E \restriction x = (x:t)\} \\
&\cup \;\{t_1(n := t_2) & &| \; t_1 \in \mathcal{DE}_E, t_2 \in \mathcal{VNF}_E\} \\
&\cup \;\{t.n & &| \; t \in \mathcal{DE}_E\} \\
&\cup \;\{t_1 \,\mathbf{where}\, n := t_2 & &| \; t_1 \in \mathcal{DE}_E, t_2 \in \mathcal{VNF}_E\} \\
&\cup \;\{\mathbf{case}\, t_1 \,\mathbf{of}\, t_2 & &| \; t_1, t_2 \in \mathcal{VNF}_E \wedge (t_1 \in \mathcal{DE}_E \vee t_2 \in \mathcal{DE}_E)\} \\
&\cup \;\{t_1 \,\mathbf{upto}\, 1 \oslash t_2 \,\mathbf{at}\, 1 & &| \; t_1, t_2 \in \mathcal{VNF}_E \wedge (t_1 \in \mathcal{DE}_E \vee t_2 \in \mathcal{DE}_E)\} \\
&\cup \;\{t_1 \,\mathbf{upto}\, m \oslash t_2 \,\mathbf{at}\, 1 & &| \; m > 1, t_2 \in \mathcal{VNF}_E \wedge t_1 \in \mathcal{DE}_E\} \\
&\cup \;\{t_1 \,\mathbf{upto}\, m \oslash t_2 \,\mathbf{at}\, n & &| \; n > 1, t_1 \in \mathcal{VNF}_E \wedge t_2 \in \mathcal{DE}_E\}
\end{aligned}
$$

It is rather straightforward to see that an expression explicitly valid under $E$ cannot be reduced any more if and only if it is a member of $\mathcal{VNF}_E$.

**Theorem 10 (Valid irreducibility).** *For any environment $E$ and expressions $e$ with $E \vdash_{val} e$ it holds that $e$ is irreducible if and only if $e \in \mathcal{VNF}_E$.*

An application of this result will be presented in the next section.

### 3.7.6 Adequacy of Formalizations

The goal of this section is to present a simple technical framework that allows to prove the adequacy of the formalization of a theory within Deva. Obviously, in order to make such a proof one has to define first what is meant by a theory. The latter is not obvious at all, in the context of this presentation the following extremely simplistic definition of a theory will be sufficient.

**Definition 11 (Theories).** A theory $T$ is a pair $\langle For, Thm \rangle$ consisting of a set of formulas *For* and the subset of theorems *Thm* of *For*.

Now assume that a theory $T$ is given and the goal is to adequately formalize that theory in Deva. Before presenting the notion of adequate formalization, one more auxiliary definition is needed.

**Definition 12 (Inhabitants set).** The inhabitants set $INH_E(t)$ of a text $t$ under a case-free environment $E$ is defined as the set of all case-free texts $t_1$ such that $E \vdash_{val} t_1 \therefore t$.

The basic idea is to formalize the theory $T$ by an environment $E_T$ that consists of an environment $E_{For}$ containing declarations producing the formulas of $T$ and an environment $E_{Ax}$ containing declarations formalizing the axioms and inference rules of $T$. The two systems are related via a bijection.

**Definition 13 (Adequate formalization).** A theory $T = \langle For, Thm \rangle$ is adequately formalized by a case-free environment $E_T = E_{For} \oplus E_{Ax}$ if and only if $\vdash_{val} E_T$ and there is a subset $T_{For}$ of $INH_{E_{For}}(\textbf{prim})$ and a bijection $\alpha : For \to T_{For}$ such that $Thm = \{ f \in For \mid INH_{E_{Ax}}(\alpha(f)) \neq \{ \} \}$.

For convenience, the primitive text **prim** has been chosen as the type of formulas.

When trying to prove adequacy of formalizations, the difficult point is usually to show that any $f \in For$ with $INH_{E_{Ax}}(\alpha(f)) \neq \{ \}$ is a member of $Thm$. A proof that transforms derivations described by the "inhabitants" of $\alpha(f)$ into derivations of $f$ inside the theory $T$ will face the problem that many such inhabitants, while syntactically completely different, are equivalent modulo conversion. However, as a consequence of confluence and strong normalization the problem can be considerably simplified, because only unique normal forms have to be considered. By using the characterization of valid forms, the following theorem allows the above problem about adequacy proofs to be simplified. It asserts that it suffices to consider those inhabitants which are also valid normal forms.

**Theorem 14 (Simplification of inhabitance test).** *For any case-free environment $E$ and text $t$ such that $E \vdash_{val} t$ it holds that $INH_E(t) \neq \{ \}$ if and only if $INH_E(t) \cap \mathcal{VNF}_E \neq \{ \}$.*

**Proof.** In order to show the interesting direction, consider a case-free text $t_1$ with $E \vdash_{val} t_1 \therefore t$: By strong normalization of case-free validity, $t_1$ has a normal form, say $t_2$. By closure of explicitly valid expressions against reduction it follows that $E \vdash_{val} t_2$. By closure of explicit validity against typing and reduction, it follows that also $E \vdash_{val} t_2 \therefore t$. By valid irreducibility, it follows that $t_2$ must be a member of $\mathcal{VNF}_E$. $\square$

A concrete example of an adequacy proof for a formalization of predicate logic and Peano arithmetic in a sublanguage of Deva can be found in [31].

## 3.8   Discussion

As could be seen, the design choices of the Deva meta-calculus lead to a lean syntax and a somewhat forbidding formal semantics. Yet, the intuitive mastery of the Deva-operations has proved relatively easy for program designers, as illustrated in several detailed case studies. Current research investigates whether this intuitive understanding could be better expressed by a algebraic semantics instead of the operational, $\lambda$-calculus based semantics underlying the current definition.

Probably the most characteristic language feature of Deva is the syntactical identification of elements and their types, following and extending the ideas of one of the AUTOMATH languages [80]. In addition, Deva offers a number of type structures for the formalization of theories and developments, e.g. named operations, case distinctions, cuts, and contexts. Contexts are lists of declarations and abbreviations together with some simple syntactic operations i.e., join, application, import, and renaming. Nevertheless, the type structure of the explicit part of Deva lacks many advanced features investigated in related type systems, e.g. polymorphism [24], dependent sums and cumulative type hierarchies [72], inclusion polymorphism [61], or subtyping [19] just to mention some.

The explanation relation defining the implicit level of Deva was designed to allow the notation and composition of proofs without having to concentrate on all the technical overhead resulting from formalization of every detail. Similar ideas are followed in [87]. When interpreted as a notation for tactics, the implicit level of Deva is much weaker than other tactic notations, as illustrated in e.g. LCF [82], HOL [45], Elf [85], or Mural [61]. For example the iteration construct **loop** is nothing but a flexible shorthand notation for repeated rule applications, in particular it does not introduce the possibility for general recursive definitions. On the other hand, there is no separate meta-notation for tactics in Deva, rather tactical constructs extend the explicit notation and inherit useful features such as the theory system, the abstraction mechanisms etc.

While the explanation of implicit expressions by explicit ones proved sufficient as a descriptive device, its strong nondeterminism precludes any naive implementation of the implicit level in full. Based on the experience gained by ongoing experimentation, current work concentrates on finding ways to guide explanation based on the information available on the implicit level. First technical results can be found in [101].

A serious problem arises from the fact that explanation is *ambiguous*, i.e., an implicit expression may allow more than one explanation. Such ambiguity is harmless when arising from the explanations of implicit proofs since it is irrelevant how the explicit proof looks like, as long as it proves the desired result. However, such ambiguity becomes extremely harmful when arising from the explanation of formulas, for quite obvious reasons. The Deva language itself does not offer any help here, the burden is on the user to discipline himself in order not to be fooled by any harmful ambiguities on the implicit level.

There remain many open points and desirables for a more complete language theory of Deva: For example, the strong normalization result should be established without the restriction have case-free expressions. Another issue concerns the inclusion of extensional properties, e.g. $\eta$-conversion into the notion of conversion in Deva. Nevertheless, a number of basic positive results have been presented and illustrated to establish a theoretical basis to perform adequacy proofs of Deva-formalizations and to develop sound support systems for Deva.

Apart from the very syntactic *term model*, in which every term is interpreted by its normal form, there is no model theory for Deva. All consistency proofs have been performed purely on the syntactic definitions. Intuitively, the main obstacle for a naive set-theoretic semantic is the fact that a hierarchy of Deva types is constructed "downwards", i.e., from types to values, as opposed to a hierarchy of typed sets which is constructed "upwards", i.e., from values to types.

# 4  Formalization of Basic Theories

In this chapter, some selected basic theories of mathematics and logic are formalized as Deva contexts. There are several reasons to present these formalizations separate from the following two main case studies: First, the power of expression, and in particular the modular formalization style, of Deva is demonstrated. Second, these formalizations are mostly concerned with standard theory and thus should be easily understandable for those readers who are not experts wrt. the technicalities of Deva. These formalizations will serve as a library for both of the two formal development case studies in the following chapters.

The formalizations given in this chapter are not meant to be definite in the sense that the theories have to be formalized in this way. In fact, many different formalizations are thinkable. The authors have chosen these particular formalizations to suit their needs and tastes. Much material has been taken or adapted from a collection of Deva-formalizations of basic theories [90].

This chapter is a self contained WEB document: It has been checked separately from the two case studies by the Devil system. Then, it has been stored in such a way that both case studies could import it, without the need to check it all over again.

## 4.1  Overview

In this and the following two chapters we will begin with an overview over what is to be formalized, so that the reader can gain some understanding of the global structure of the formalization. The chapter consists of three parts: a logical basis and basic theories for the two case studies.

**context** *BasicTheories* :=
⟦⟨ Logical basis. 4.1.1 ⟩
; ⟨ Basic theories of VDM. 4.1.2 ⟩
; ⟨ Basic theories for algorithm calculation. 4.1.3 ⟩
⟧

**4.1.1.**  We begin with a formalization of a suitable logical basis: predicate logic and, based on it, two notions of equality. As in the tutorial chapter, *prop* denotes the type of propositions and *sort* denotes the type of sort-variables.

⟨ Logical basis. 4.1.1 ⟩ ≡
**context** *LogicalBasis* :=
⟦ *prop*, *sort* : **prim**
; ⟨ Propositional Logic. 4.2.1.1 ⟩
; ⟨ Predicate Logic. 4.2.1.7 ⟩
; ⟨ Parametric equality of terms. 4.2.2 ⟩

; ⟨ Parametric equality of functions. 4.2.3 ⟩

]

This code is used in section 4.1.


**4.1.2.**   In Sect. 4.3, some theories and datastructures which are essential for the VDM reification case study are formalized. Included are classical first-order sorted logic with (nonextensional) equality, natural numbers, sets, sequences, tuples, and finite maps. And finally, a small library of tactics specific to the VDM datastructures is included.

⟨ Basic theories of VDM. 4.1.2 ⟩ ≡

**context**   *VDMBasics* :=

[ ⟨ Imports needed by *VDMBasics*. 4.3 ⟩

; ⟨ Natural numbers. 4.3.1 ⟩

; ⟨ Finite sets. 4.3.2 ⟩

; ⟨ Sequences. 4.3.3 ⟩

; ⟨ Tuples. 4.3.4 ⟩

; ⟨ Finite maps. 4.3.5 ⟩

; ⟨ VDM tactics. 4.3.6 ⟩

]

This code is used in section 4.1.

Some difficulties arise when reasoning in VDM about *partial* functions. A non-standard logic coping with these problems is described in [11]. This logic is known as the *Logic of Partial Functions* (LPF). It makes use of a third constant, besides *true* and *false*, for denoting missing values. The main differences between LPF and predicate logic (which is formalized below) are that

   −  the rule of excluded middle does not hold, and that
   −  the deduction theorem does not hold without an additional hypothesis about the "definedness" of the proposition in the premise.

The rules of LPF (derived from [65]) could have easily been translated into Deva. Nevertheless, for clarity and simplicity, LPF and the problems it raises have been discarded in the context of this case study. For example, including a formalization of LPF would have the effect of significantly complicating formal developments in VDM because of the necessity of a significant amount of additional proofs verifying definedness hypotheses.

**4.1.3.**   In Sect. 4.4, some theories essential for the case study on algorithm calculation are formalized. They consist of classical first-order sorted logic with extensional equality and some elementary notions from algebra.

⟨ Basic theories for algorithm calculation. 4.1.3 ⟩ ≡

**context** *CalculationalBasics* :=

⟦ ⟨ Imports needed by *CalculationalBasics*. 4.4.1 ⟩

; ⟨ Extensional equality of terms or functions. 4.4.1.1 ⟩

; ⟨ Terms involving functions. 4.4.2.1 ⟩

; ⟨ Some Bits of Algebra. 4.4.3.1 ⟩

; ⟨ Induced Partial Ordering. 4.4.4 ⟩

⟧

This code is used in section 4.1.

Adequacy of the formalizations will not be proven, i.e., the burden is on the reader to check whether the syntactic objects presented in this chapter are adequate formalizations of the corresponding theories. Some hints on how to perform such adequacy proofs have been given in Sect. 3.7.6.

## 4.2  Logical Basis

### 4.2.1  Classical Many-Sorted Logic

**4.2.1.1.**  First, the classical propositional calculus is formalized in a natural deduction style similar to [88].

⟨ Propositional Logic. 4.2.1.1 ⟩ ≡

**context** *PropositionalLogic* :=

⟦ ⟨ Operators of propositional logic. 4.2.1.2 ⟩

; ⟨ Axioms of propositional logic. 4.2.1.3 ⟩

; ⟨ Derived operators of propositional logic. 4.2.1.4 ⟩

; ⟨ Law of the excluded middle. 4.2.1.5 ⟩

; ⟨ Derived laws of propositional logic. 4.2.1.6 ⟩

⟧

This code is used in section 4.1.1.

**4.2.1.2.**  Here, we introduce the basic propositional constant *false* and the logical connectives.

⟨ Operators of propositional logic. 4.2.1.2 ⟩ ≡

$\quad$ *false* $\quad$ : *prop*

; $(\cdot) \Rightarrow (\cdot)$,

$(\cdot) \wedge (\cdot),$

$(\cdot) \vee (\cdot) \quad : [\,prop;\, prop \vdash prop\,]$

This code is used in section 4.2.1.1.


**4.2.1.3.** Here we state the introduction and elimination laws for the operators introduced in the previous section.

$\langle$ Axioms of propositional logic. 4.2.1.3 $\rangle \equiv$

$$false\_elim \;:\; [\,p\,?\,prop \;\vdash\; \left|\dfrac{false}{p}\right.]$$

$$;\; imp \qquad :\; [\,p, q\,?\,prop \;\vdash\; \langle\, intro := \left|\dfrac{[p \vdash q]}{p \Rightarrow q}\right., \;\; elim := \left|\dfrac{p;\, p \Rightarrow q}{q}\right.\rangle\,]$$

$$;\; conj \qquad :\; [\,p, q\,?\,prop \;\vdash\; \langle\, intro := \left|\dfrac{p;\, q}{p \wedge q}\right.\right.$$

$$, \;\; eliml := \left|\dfrac{p \wedge q}{p}\right., \;\; elimr := \left|\dfrac{p \wedge q}{q}\right.\rangle$$

$$]$$

$$;\; disj \qquad :\; [\,p, q\,?\,prop \;\vdash\; \langle\, introl := \left|\dfrac{p}{p \vee q}\right., \;\; intror := \left|\dfrac{q}{p \vee q}\right.\right.$$

$$, \;\; elim \;\; := [\,r\,?\,prop \;\vdash\; \left|\dfrac{p \vee q;\, [p \vdash r];\, [q \vdash r]}{r}\right.]$$

$$\rangle$$

$$]$$

This code is used in section 4.2.1.1.


**4.2.1.4.** Negation, equivalence, and the constant *true* can be derived from the operators already defined. It is also helpful to define the notion of an identity of propositions (cf. Chap. 5).

$\langle$ Derived operators of propositional logic. 4.2.1.4 $\rangle \equiv$

$\neg\,(\cdot) \qquad := [\,p : prop \vdash p \Rightarrow false\,]$

$;\; (\cdot) \Leftrightarrow (\cdot) := [\,p, q : prop \vdash (p \Rightarrow q) \wedge (q \Rightarrow p)]$

$;\; true \qquad := \neg\; false$

$;\; prop\_id := [\,p\,?\,prop \vdash [\,u : p \vdash u\,]]$

This code is used in section 4.2.1.1.

**4.2.1.5.** By adding the law of the excluded middle one passes from the *intuitionistic* propositional calculus to the *classical* propositional calculus.

⟨ Law of the excluded middle. 4.2.1.5 ⟩ ≡
    $excluded\_middle$ : $[\, p\, ?\, prop \vdash p \vee \neg\ p\, ]$

This code is used in section 4.2.1.1.

**4.2.1.6.** As a simple illustration of the axioms, consider the following proof of the proposition *true*.

⟨ Derived laws of propositional logic. 4.2.1.6 ⟩ ≡
    $true\_proof$ := $imp$ . $intro$ $([\, x : false \vdash x\, ])$ ∴ $true$

This code is used in section 4.2.1.1.

**4.2.1.7.** The predicate calculus is obtained by adding laws for the two quantifiers in a style again adopted from [88].

⟨ Predicate Logic. 4.2.1.7 ⟩ ≡
**import context** *PredicateLogic* :=
⟦ **import** *PropositionalLogic*
; $\forall(\cdot), \exists(\cdot)$ : $[\, s\, ?\, sort\, ; [\, s \vdash prop\, ] \vdash prop\, ]$
; *univ*      :
    $[\, s\, ?\, sort\, ;\, P\, ?\, [\, s \vdash prop\, ]$
    $\vdash\!\!\langle\ intro$ := $\dfrac{[\, x : s \vdash P(x)\, ]}{\forall\, P}$, $elim$ := $\dfrac{x\ :\ s\, ;\, \forall\, P}{P\, (x)}\!\rangle$
    $]$
; *ex*      :
    $[\, s\, ?\, sort\, ;\, P\, ?\, [\, s \vdash prop\, ]$
    $\vdash\!\!\langle\ intro$ := $\dfrac{x\, ?\, s\, ;\, P(x)}{\exists\, P}$, $elim$ := $[\, p\, ?\, prop\, \vdash \dfrac{[\, x : s\, ;\, P(x) \vdash p\, ];\, \exists\, P}{p}\, ]\rangle$
    $]$
; ⟨ Further axioms and laws of predicate logic. 4.2.1.8 ⟩
⟧

This code is used in section 4.1.1.

Note that quantification is expressed in a slightly modified notation. For example, the formula $\forall x : nat.p(x)$ is expressed by $\forall[\, x : nat \vdash p(x)\, ]$.

**4.2.1.8.** While in principle these axioms provide all the power of predicate logic, they are not very effective to use in concrete proof situations. For example, one often wants to substitute equivalent formulas for each other inside another formula, but this process becomes rather tedious if one is restricted to use only the preceding axioms. For this reason, a deduction principle is added as an axiom, stating that equivalent propositions may be substituted within a proposition without changing its validity. Note that this principle is at the meta-level wrt. predicate calculus, and thus, it cannot be derived from the given axioms without some laws about the meta-level (e.g. a principle of structural induction on proposition schemas)

⟨ Further axioms and laws of predicate logic. 4.2.1.8 ⟩ ≡
$psubst$ : [ $S$ ?[$prop$ ⊢ $prop$]; $p, q$ ? $prop$ ⊢ [[$p$ ⊨ $q$] ⊢ [ $S$ ($p$) ⊢ $S(q)$]]]

See also sections 4.2.1.9, 4.2.1.10, 4.2.1.11, and 4.2.1.12.

This code is used in section 4.2.1.7.


We have introduced a new notation to denote equivalence between propositions $p$ and $q$. This notation is shorthand for a binary product describing mutual derivability between $p$ and $q$:

$$[p \vDash q] \quad \equiv \quad ⟨\ down := [p \vdash q]\ , up := [q \vdash p]\ ⟩$$

There is also a two dimensional variant of this notation. From here on, this shorthand notation will be used throughout the book.

**4.2.1.9.** As an illustration of how to use these axioms we present certain rules and laws which we will use in the subsequent chapters. We will state these laws without supplying explicit proof texts for them. However, it is an instructive exercise for the reader to fill in the gaps by finding texts of the appropriate type. First, the symmetric version of the preceding deduction principle is denoted by *prsubst*:

⟨ Further axioms and laws of predicate logic. 4.2.1.8 ⟩+ ≡
; $prsubst$ : [ $S$ ?[$prop$ ⊢ $prop$]; $p, q$ ? $prop$ ⊢ [[$p$ ⊨ $q$] ⊢ [ $S$ ($q$) ⊢ $S(p)$]]]


**4.2.1.10.** *equiv_props* combines some useful laws concerning equivalence, and similar *conj_props* and *disj_props*.

⟨ Further axioms and laws of predicate logic. 4.2.1.8 ⟩+ ≡
; *conj_props* :

$$\langle\ commut := [p, q\ ?\ prop \vdash \frac{p \wedge q}{q \wedge p}]$$

$$,\ simpl\quad := [\,p\ ?\ prop \vdash \langle\ false := \frac{false \wedge p}{false},\ true := \frac{true \wedge p}{p}\,\rangle\,]$$

$$,\ simpr\quad := [\,p\ ?\ prop \vdash \langle\ false := \frac{p \wedge false}{false},\ true := \frac{p \wedge true}{p}\,\rangle\,]$$

$$\rangle$$

$;\ disj\_props\quad :$

$$\langle\ commut := [p, q\ ?\ prop \vdash \frac{p \vee q}{q \vee p}]$$

$$,\ simpl\quad := [\,p\ ?\ prop \vdash \langle\ false := \frac{false \vee p}{p},\ true := \frac{true \vee p}{true}\,\rangle\,]$$

$$,\ simpr\quad := [\,p\ ?\ prop \vdash \langle\ false := \frac{p \vee false}{p},\ true := \frac{p \vee true}{true}\,\rangle\,]$$

$$\rangle$$

$;\ equiv\_props\ :$

$$\langle\ refl\qquad := [p, q\ ?\ prop \vdash \frac{p \Leftrightarrow p}{true}]$$

$$,\ sym\qquad := [p, q\ ?\ prop \vdash \frac{p \Leftrightarrow q}{q \Leftrightarrow p}]$$

$$,\ trans\qquad := [p, q, r\ ?\ prop \vdash \frac{p \Leftrightarrow q;\ q \Leftrightarrow r}{p \Leftrightarrow r}]$$

$$,\ conj\_right := [p, q, r\ ?\ prop \vdash [r \vdash \frac{p \Leftrightarrow q}{p \Leftrightarrow q \wedge r}]]$$

$$,\ mon\_orl\quad := [p, q, r\ ?\ prop \vdash \frac{p \Leftrightarrow q}{p \vee r \Leftrightarrow q \vee r}]$$

$$,\ mon\_conjl := [p, q, r\ ?\ prop \vdash [r \vdash \frac{p \Leftrightarrow q}{p \wedge r \Leftrightarrow q \wedge r}]]$$

$$,\ inj\_fl\qquad := [\,p\ ?\ prop \vdash [\neg\ p \vdash [p \vDash false]]]$$

$$,\ switch\qquad := [p, q\ ?\ prop \vdash \frac{p \Leftrightarrow q}{[p \vDash q]}]$$

$$\rangle$$

**4.2.1.11.** We can also define abbreviations for more complex operations which are constructed from the atomic ones. For example, we will make use of an existential quantification of a binary predicate.

⟨ Further axioms and laws of predicate logic. 4.2.1.8 ⟩+ ≡

$; \exists_2 (\cdot) := [a, b \, ? \ \ sort \ \vdash [ \, P :[a; b \vdash prop \, ] \vdash \exists [ \, x : a \vdash \exists \, P \, (x) ] ] ]$

**4.2.1.12.** In subsequent chapters, we will repeatedly use some of the above laws in order to simplify a formula. Hence, for economy, we define the alternative **alt** between these laws as a simplification tactic:

⟨Further axioms and laws of predicate logic. 4.2.1.8⟩+ ≡
; *LogicalSimplification* :=
   **alt** [ *psubst*( *conj_props. simpl .true*) , *psubst*( *conj_props. simpl .false*)
      , *psubst*( *conj_props. simpr .true*), *psubst*( *conj_props. simpr .false*)
      , *psubst*( *disj_props. simpl .true*) , *psubst*( *dis . props. simpl .false*)
      , *psubst*( *disj_props. simpr .true*) , *psubst*( *disj . props. simpr .false*)
      , *psubst*( *equiv_props.refl*)
      ]

Note the use of the substitution axiom *psubst* whicl allows to simplify not only at the top-level but anywhere inside an expression.

    This concludes the formalization of classical first-order logic.

### 4.2.2 Parametric Equality of Terms

Parametric term equality is formalized by stating properties for reflexivity, substitution, symmetry, and transitivity. The first two of these properties are stated as axioms from which the other properties can be derived.

⟨Parametric equality of terms. 4.2.2⟩ ≡
**context** *EqualityOfTerms* :=
$[\![ \ (\cdot) = (\cdot) \ : [ \, s \, ? \ sort \ \vdash [s; s \vdash prop \, ] ]$
$; \ refl \qquad : [ \, s \, ? \ sort \, ; x \, ? \ s \vdash x = x \, ]$
$; \ subst \quad \ : [ \, s \, ? \ sort \, ; x, y \, ? \ s \, ; P \, ? \, [s \vdash prop \, ] \vdash [x = y \vdash \left| \dfrac{P \, (x)}{P \, (y)} \right| ] ]$
$; ⟨\text{Properties of parametric equality of terms. 4.2.2.1}⟩$
$]\!]$

This code is used in section 4.1.1.

Note that the axiom *subst* distinguishes equality from an ordinary equivalence relation.

**4.2.2.1.** Symmetry can be derived by suitably instantiating the substitution property and then cleaning up the resulting formula using reflexivity:

⟨Properties of parametric equality of terms. 4.2.2.1⟩ ≡

$$sym := subst\,(\mathbf{6} := refl) \therefore [\,s\,?\,sort\,;x,y\,?\ \ s\ \vdash \left|\begin{array}{c} x = y \\ \hline y = x \end{array}\right.]$$

See also sections 4.2.2.2, 4.2.2.3, 4.2.2.4, 4.2.2.5, and 4.2.2.6.

This code is used in section 4.2.2.

**4.2.2.2.** As a simple exercise, the reader is invited to derive a proof text for inequality.

⟨Properties of parametric equality of terms. 4.2.2.1⟩+ ≡
; $symneg$ : [ $s$ ? $sort$ ; $x, y$ ? $s \vdash [\neg\, x = y \vdash \neg\, y = x\,]]$

**4.2.2.3.** Transitivity of equality can be derived by suitably instantiating the substitution property and then cutting with the symmetry property just derived.

⟨Properties of parametric equality of terms. 4.2.2.1⟩+ ≡
; $trans$ := [ $s$ ? $sort$ ; $z$ ? $s$
     $\vdash subst\,(P := [\,w : s \vdash w = z\,])\ \Diamond\!\!\!\!\!\!\!\circ\ sym$
     ]

$$\therefore [\,s\,?\,sort\,;x,y,z\,?\ \ s\ \vdash \left|\begin{array}{c} x = y;\, y = z \\ \hline x = z \end{array}\right.]$$

**4.2.2.4.** Substitution can be suitably specialized to derive the unfold rule.

⟨Properties of parametric equality of terms. 4.2.2.1⟩+ ≡
; $unfold$ := [$s, t$ ? $sort$ ; $z$ ? $t$ ; $F$ ? [$s \vdash t$]
      $\vdash subst\,(P := [\,w : s \vdash z = F(w)\,])$
      ]

$$\therefore [s, t\,?\ sort\,;x,y\,?\ s\,;z\,?\ t\,;F\,?\ [s \vdash t]\vdash [x = y \vdash \left|\begin{array}{c} z = F(x) \\ \hline z = F(y) \end{array}\right.]]$$

**4.2.2.5.** By cutting both substitution and the unfold rule with symmetry, one obtains reversed substitution and the fold rule.

⟨Properties of parametric equality of terms. 4.2.2.1⟩+ ≡
; $rsubst$ := $sym \Diamond\!\!\!\!\!\!\!\circ subst$

$$\therefore [\,s\,?\,sort\,;x,y\,?\ s\,;P\,?\ [s \vdash prop\,]\vdash [x = y \vdash \left|\begin{array}{c} P\,(y) \\ \hline P\,(x) \end{array}\right.]]$$

; *fold*    := *sym* ⋄ *unfold*

$$\therefore [s, t\,?\ sort\,;x, y\,?\ s\,;z\,?\ t\,;F\,?\,[s \vdash t]\vdash [x = y \vdash \begin{vmatrix} z = F(y) \\ z = F(x) \end{vmatrix}]]$$

**4.2.2.6.** An example for the application of the reversed substutivity is the following simple law which could have equally been derived by an application of the fold rule, followed by an application of the symmetry rule. Simply by stating the desired type of the expression *rsubst* ($hyp_1, hyp_2$) the machine is enabled to deduce the predicate $P$. This demonstrates the power of the meta-calculus and it shows how much the user is freed from stating boring detail.

⟨ Properties of parametric equality of terms. 4.2.2.1 ⟩+ ≡
; *eq_compose* :=
$[s, t, u\,?\ sort\,;\ a\,?\ s\,;\ b\,?\ t\,;\ c\,?\ u\,;\ f\,?\,[s \vdash t];\ g\,?\,[t \vdash u]$
$$\left|\begin{array}{l} hyp_1\ :\ f\,(a) = b; \\ hyp_2\ :\ g\,(b) = c \\ \hline rsubst\,(hyp_1, hyp_2) \therefore g(f(a)) = c \end{array}\right.$$
$]$

$$\therefore [s, t, u\,?\ sort\,;\ a\,?\ s\,;\ b\,?\ t\,;\ c\,?\ u\,;\ f\,?\,[s \vdash t];\ g\,?\,[t \vdash u]$$
$$\left|\begin{array}{c} f\,(a) = b; g(b) = c \\ \hline g\,(f(a)) = c \end{array}\right.$$
$]$

### 4.2.3   Parametric Equality of Functions

In the case study on algorithm calculation (Chap. 6), it is not sufficient to just consider terms of some sort $s$, but we also have to take functions, i.e., objects of a functional type $[s \vdash t]$, into account. Since that case study mainly relies on equational reasoning, we need an additional equality of functions, because the equality just introduced is not applicable to such objects. A theory of equality of functions can be formalized by replacing terms by functions in the theory of equality of terms presented above. The signature of the equality relation and the axioms of reflexivity and substitution can be adapted without problems. Note that, in the logic of partial functions, neither reflexivity nor substitution hold. A stronger equality has to be introduced and it is a good exercise to try formalize such an equality.

⟨ Parametric equality of functions. 4.2.3 ⟩ ≡
**context** *EqualityOfFunctions* :=
$[\!\![\ (\cdot) = (\cdot)\ :\ [s, t\,?\ sort\ \vdash\ [[s \vdash t]; [s \vdash t]\ \vdash\ prop\,]]$
; *refl*      $:\ [s, t\,?\ sort\,;f\,?\,[s \vdash t]\vdash f = f]$

$$; \; subst \quad : [s, t \,?\; sort \,; f, g \,?\; [s \vdash t]; P \,?\; [[s \vdash t] \vdash prop] \vdash [f = g \vdash \left|\frac{P\,(f)}{P\,(g)}\right|]]$$

$; \langle$ Properties of parametric equality of functions. 4.2.3.1 $\rangle$

]

This code is used in section 4.1.1.

**4.2.3.1.** As in the theory of equality of terms, other rules and properties can be derived by adapting the corresponding proofs.

$\langle$ Properties of parametric equality of functions. 4.2.3.1 $\rangle \equiv$

$$sym \quad : [s, t \,?\; sort \,; f, g \,?\; [s \vdash t] \vdash \left|\frac{f = g}{g = f}\right|]$$

$$; \; trans \quad : [s, t \,?\; sort \,; f, g, h \,?\; [s \vdash t] \vdash \left|\frac{f = g; g = h}{f = h}\right|]$$

$; \; unfold : [s, t, u, v \,?\; sort \,; f, g \,?\; [s \vdash t]; h \,?\; [u \vdash v]; F \,?\; [[s \vdash t] \vdash [u \vdash v]]$

$$\vdash [f = g \vdash \left|\frac{h = F(f)}{h = F(g)}\right|]$$

]

$; \; rsubst \quad : [s, t \,?\; sort \,; f, g \,?\; [s \vdash t]; P \,?\; [[s \vdash t] \vdash prop]$

$$\vdash [g = f \vdash \left|\frac{P\,(f)}{P\,(g)}\right|]$$

]

$; \; fold \quad : [s, t, u, v \,?\; sort \,; f, g \,?\; [s \vdash t]; h \,?\; [u \vdash v]; F \,?\; [[s \vdash t] \vdash [u \vdash v]]$

$$\vdash [f = g \vdash \left|\frac{h = F(g)}{h = F(f)}\right|]$$

]

This code is used in section 4.2.3.

## 4.3   Basic Theories of VDM

The formalization of the basic theories of VDM will make use of the logical basis and the term equality which were defined in the previous section.

$\langle$ Imports needed by *VDMBasics.* 4.3 $\rangle \equiv$

**import**   *LogicalBasis* [*EqualityOfTerms* =: *Equality* ]

$;$ **import**   *Equality*

This code is used in section 4.1.2.

### 4.3.1 Natural Numbers

VDM provides the built-in types $\mathbb{B}$, $\mathbb{N}$, $\mathbb{N}_1$, $\mathbb{Z}$, and $\mathbb{R}$ but only $\mathbb{N}$ will be formalized here. The formalization is straightforward (cf. Sect. 2.1.5) and should not require further explanation.

⟨ Natural numbers. 4.3.1 ⟩ ≡

**import context** *NaturalNumbers* :=

⟦ *nat* : *sort*

; ⟨ Constructors of natural numbers. 4.3.1.1 ⟩

; ⟨ Axioms of natural numbers. 4.3.1.2 ⟩

; ⟨ Derived operators of natural numbers. 4.3.1.3 ⟩

⟧

This code is used in section 4.1.2.

**4.3.1.1.**  ⟨ Constructors of natural numbers. 4.3.1.1 ⟩ ≡

0      : *nat*

; *succ* : [*nat* ⊢ *nat*]

This code is used in section 4.3.1.

**4.3.1.2.**  ⟨ Axioms of natural numbers. 4.3.1.2 ⟩ ≡

   *peano*$_3$     : [ $x$ ? *nat* ⊢ ¬ *succ* $(x) = 0$]

; *peano*$_4$     : [$x, y$ ? *nat* ⊢ $\dfrac{succ\,(x) = succ(y)}{x = y}$]

; *induction* : [ $P$ ?[*nat* ⊢ *prop*] ⊢ $\dfrac{P\,(0);\,[\,x\,?\,nat\,;\,P(x) \vdash P(succ(x))\,]}{[\,x\,?\,nat \vdash P(x)\,]}$]

This code is used in section 4.3.1.

**4.3.1.3.**  ⟨ Derived operators of natural numbers. 4.3.1.3 ⟩ ≡

1          := *succ* $(0)$

; 2          := *succ* $(1)$

; $(\cdot) + (\cdot)$  : [*nat*; *nat* ⊢ *nat*]

; *add_def*  : ⟨ *base* := [ $m$ ? *nat* ⊢ $0 + m = m$]

              , *rec*  := [$n, m$ ? *nat* ⊢ $succ(n) + m = succ(n + m)$]

              ⟩

; *pred*   : [*nat* ⊢ *nat*]

; $pred\_def$ : $[\, n \,?\, nat \,\vdash\, pred(succ(n)) = n \,]$

; $(\cdot) - (\cdot)$ : $[\, nat;\, nat \vdash nat \,]$

; $sub\_def$ : $\langle\, base := [\, m \,?\, nat \,\vdash\, m - 0 = m \,]$

  , $rec$ := $[\, n, m \,?\, nat \,\vdash\, succ(n) - succ(m) = n - m \,]$

  $\rangle$

; $(\cdot) \times (\cdot)$ : $[\, nat;\, nat \vdash nat \,]$

; $mult\_def$ : $\langle\, base := [\, n \,?\, nat \,\vdash\, 0 \times n = 0 \,]$

  , $rec$ := $[\, n, m \,?\, nat \,\vdash\, succ(n) \times m = n \times m + m \,]$

  $\rangle$

; $(\cdot)^{(\cdot)}$ : $[\, nat;\, nat \vdash nat \,]$

; $exp\_def$ : $\langle\, base := [\, n \,?\, nat \,\vdash\, n^0 = 1 \,]$

  , $rec$ := $[\, n, m \,?\, nat \,\vdash\, n^{succ(m)} = n^m \times n \,]$

  $\rangle$

; $(\cdot) \leq (\cdot)$ : $[\, nat;\, nat \vdash prop \,]$

; $leq\_def$ : $\langle\, base := [\, n \,?\, nat \,\vdash\, 0 \leq n \,]$

  , $rec$ := $[\, n, m \,?\, nat \,\vdash\, [\, n \leq m \vdash succ(n) \leq succ(m) \,] \,]$

  $\rangle$

; $(\cdot) > (\cdot)$ := $[\, n, m : nat \,\vdash\, \neg\, n \leq m \,]$

; $(\cdot) \geq (\cdot)$ := $[\, n, m : nat \,\vdash\, n > m \vee n = m \,]$

; $(\cdot) < (\cdot)$ := $[\, n, m : nat \,\vdash\, \neg\, n \geq m \,]$

This code is used in section 4.3.1.

In addition to these operations, VDM has further operations on naturals, e.g. *rem*, *mod*, and *div*, with obvious meaning. More such operations could be added but the current selection will suffice for our purposes. The issue of partiality arises when trying to define functions like the predecessor function on naturals. Some logic of partial functions would be needed as a basic theory to define such functions more adequately.

### 4.3.2 Finite Sets

$\langle$ Finite sets. 4.3.2 $\rangle \equiv$

**import context** *FiniteSets* :=

$[\!\![$ *set* : $[\, sort \vdash sort \,]$

; $\langle$ Constructors of finite sets. 4.3.2.1 $\rangle$

; $\langle$ Axioms of finite sets. 4.3.2.2 $\rangle$

; $\langle$ Derived operators and laws of finite sets. 4.3.2.3 $\rangle$

$]\!\!]$

This code is used in section 4.1.2.

**4.3.2.1.**   Finite sets in VDM are formalized from the empty set {} and the constructor $\odot$ which inserts an element into a set.

⟨ Constructors of finite sets. 4.3.2.1 ⟩ ≡
  {}        : [ $s$ ? $sort$ ⊢ $set(s)$ ]
; $(\cdot) \odot (\cdot)$ : [ $s$ ? $sort$ ⊢ [$s; set(s)$ ⊢ $set(s)$]]

This code is used in section 4.3.2.

**4.3.2.2.**   Sets do not contain duplicate elements and the order of insertion is immaterial.

⟨ Axioms of finite sets. 4.3.2.2 ⟩ ≡
  $set\_absorp$   : [ $s$ ? $sort$ ; $a$ ? $s$ ; $A$ ? $set(s)$ ⊢ $a \odot (a \odot A) = a \odot A$ ]
; $set\_commut$   : [ $s$ ? $sort$ ; $a, b$ ? $s$ ; $A$ ? $set(s)$
                  ⊢ $a \odot (b \odot A) = b \odot (a \odot A)$
                  ]
; $set\_induction$ : [ $s$ ? $sort$ ; $P$ ? [ $set(s)$ ⊢ $prop$ ]

$$\vdash \frac{P(\{\}); [\, e : s\, ; x : \ set(s)\, \vdash \dfrac{P(x)}{P(e \odot x)}\,]}{[\, x\ ?\ set(s) \vdash P(x)\,]}$$

                  ]

This code is used in section 4.3.2.

**4.3.2.3.**   A finite set can be filtered by a predicate. This mechanism will be used to mimic set comprehension.

⟨ Derived operators and laws of finite sets. 4.3.2.3 ⟩ ≡
  $(\cdot) \rhd (\cdot)$ : [ $s$ ? $sort$ ⊢ [[$s$ ⊢ $prop$]; $set(s)$ ⊢ $set(s)$]]
; $def_\rhd$     : [ $s$ ? $sort$ ; $P$ ? [$s$ ⊢ $prop$]
            ⊢⟨ $base$ := $P \rhd \{\} = \{\}$
              , $rec$   := [ $a$ ? $s$ ; $A$ ? $set(s)$
                      ⊢⟨ $accept$ := [ $P(a)$ ⊢ $P \rhd a \odot A = a \odot (P \rhd A)$ ]
                        , $reject$ := [¬ $P(a)$ ⊢ $P \rhd a \odot A = P \rhd A$ ]
                        ⟩
                      ]
              ⟩
            ]

See also sections 4.3.2.4, 4.3.2.5, 4.3.2.6, 4.3.2.7, 4.3.2.8, and 4.3.2.9.

This code is used in section 4.3.2.

Thus the notation $P \rhd A$ is equivalent to the set comprehension notation $\{a \in A \mid P(a)\}$ used in VDM. We have chosen a more succinct notation, because it leads to more succinct proofs.

**4.3.2.4.** As a typical application of the filtering operation, consider the following axioms defining the cardinality of a set.

$\langle$ Derived operators and laws of finite sets. 4.3.2.3 $\rangle+\equiv$
; $card$     : $[\, s\,?\, sort \,\vdash\, [\, set\,(s)\, \vdash\, nat\,]\,]$
; $card\_def$ : $\langle\!\langle\, base := card\,(\{\}) = 0$
            , $rec$   $:= [\, s\,?\, sort \,;\, a\,?\, s\,;\, A\,?\, set\,(s)$
                    $\vdash card\,(a \odot A) = 1 + card([\, b : s \,\vdash\, \neg\, b = a\,] \rhd A)$
               $]$
    $\rangle\!\rangle$

**4.3.2.5.** Set membership is recursively defined as follows:

$\langle$ Derived operators and laws of finite sets. 4.3.2.3 $\rangle+\equiv$
; $(\cdot) \in (\cdot)$ :  $[\, s\,?\, sort \,\vdash\, [s;\, set(s) \,\vdash\, prop\,]\,]$
; $(\cdot) \notin (\cdot) := [\, s\,?\, sort \,\vdash\, [\, a : s \,;\, A : \, set\,(s) \,\vdash\, \neg\, a \in A\,]\,]$
; $def_\in$    : $[\, s\,?\, sort \,;\, a\,?\, s$

       $\vdash\!\langle\!\langle\, base := a \notin \{\}, rec := [\, b\,?\, s \,;\, A\,?\, set\,(s) \,\vdash\, \dfrac{a \in b \odot A}{a = b \vee a \in A}\,]\rangle\!\rangle$

     $]$

**4.3.2.6.** The observability property states that two sets are equal if and only if all their elements are equal. We omit the inductive proof.

$\langle$ Derived operators and laws of finite sets. 4.3.2.3 $\rangle+\equiv$

; $observ$ : $[\, s\,?\, sort \,;\, A, B\,?\, set\,(s) \,\vdash\, \dfrac{A = B}{\forall[\, a : s \,\vdash\, a \in A \Leftrightarrow a \in B\,]}\,]$

**4.3.2.7.** The set map function $*$ applies a given function to all the elements of a given set.

$\langle$ Derived operators and laws of finite sets. 4.3.2.3 $\rangle+\equiv$

$; (\cdot) * (\cdot) : [s, t ? \ sort \vdash [[s \vdash t]; set(s) \vdash set(t)]]$
$; def_* \quad : [s, t ? \ sort ; f \ ? [s \vdash t]$
$\qquad \vdash\!\langle base := f * \{\} = \{\}$
$\qquad , \ rec \ := [\, a ? s ; A ? \ set\,(s) \vdash f * (a \odot A) = f(a) \odot (f * A)\,]$
$\qquad \quad \rangle$
$\qquad ]$

**4.3.2.8.**  All the familiar set operators are now obtained by definition, except the union which can not be expressed as a restriction.

$\langle$ Derived operators and laws of finite sets. 4.3.2.3 $\rangle + \equiv$
$; (\cdot) \cup (\cdot) \ : \ [\,s ? sort \vdash [\, set\,(s); set(s) \vdash set(s)]\,]$
$; def_\cup \qquad : \ [\,s \ ? \ sort ; A \ ? \ set\,(s)$
$\qquad \vdash\!\langle base := \{\} \cup A = A$
$\qquad , \ rec \ := [\, a ? s ; B ? \ set\,(s) \vdash (a \odot A) \cup B = a \odot (A \cup B)]$
$\qquad \quad \rangle$
$\qquad ]$
$; (\cdot) \cap (\cdot) \ := [\,s ? sort \vdash [A, B : \ set\,(s) \vdash [\, a : s \vdash a \in B\,] \rhd A]\,]$
$; (\cdot) \backslash (\cdot) \ := [\,s ? sort \vdash [A, B : \ set\,(s) \vdash [\, a : s \vdash a \notin B\,] \rhd A]\,]$
$; (\cdot) \subseteq (\cdot) := [\,s ? sort \vdash [A, B : \ set\,(s) \vdash \forall[\, a : s \vdash a \in A \Rightarrow a \in B]]\,]$

**4.3.2.9.**  Many properties about sets could be derived from these axioms. One such property are listed without proof as typical examples.

$\langle$ Derived operators and laws of finite sets. 4.3.2.3 $\rangle + \equiv$
$; props\_in : [\,s \ ? \ sort ; a \ ? \ s ; A \ ? \ set\,(s)$

$$\vdash\!\langle filter \ := [\, P ? [s \vdash prop\,] \vdash \frac{a \in P \rhd A}{a \in A \wedge P(a)}]$$

$$, \ union := [\, B ? \ set\,(s) \vdash \frac{a \in A \cup B}{a \in A \vee a \in B}]$$

$\qquad \quad \rangle$
$\qquad ]$

## 4.3.3  Sequences

$\langle$ Sequences. 4.3.3 $\rangle \equiv$
**import context** *Sequences* :=
$[\![ \ seq \ : [sort \vdash sort\,]$

; ⟨ Constructors of sequences. 4.3.3.1 ⟩
; ⟨ Axioms of sequences. 4.3.3.2 ⟩
; ⟨ Derived operators and laws of sequences. 4.3.3.3 ⟩
]

This code is used in section 4.1.2.

**4.3.3.1.**   Sequences can be described in very much the same style as sets.

⟨ Constructors of sequences. 4.3.3.1 ⟩ ≡

$\langle \rangle$        $: [\, s\,?\, sort \vdash seq(s) \,]$
$;\, (\cdot) :: (\cdot) : [\, s\,?\, sort \vdash [s;\, seq(s) \vdash seq(s)] \,]$

This code is used in section 4.3.3.

**4.3.3.2.**   Note that in contrast to sets, there are no equations between constructors.

⟨ Axioms of sequences. 4.3.3.2 ⟩ ≡

$seq\_free_1$        $: [\, s\,?\, sort\,;\, u\,?\, s\,;\, x\,?\, seq\,(s) \vdash \neg\, u :: x = \langle \rangle \,]$

$;\, seq\_free_2$        $: [\, s\,?\, sort\,;\, u, v\,?\, s\,;\, x, y\,?\, seq\,(s) \vdash \dfrac{u :: x = v :: y}{u = v \wedge x = y} ]$

$;\, seq\_induction : [\, s\,?\, sort\,;\, P\,?\, [\, seq\,(s) \vdash prop\,]$

$$\vdash \dfrac{\dfrac{P\,(\langle \rangle);\, [\, u : s\,;\, x : seq\,(s) \vdash \dfrac{P\,(x)}{P\,(u :: x)} ]}{}}{[\, x\,?\, seq\,(s) \vdash P(x)\,]}$$

]

This code is used in section 4.3.3.

**4.3.3.3.**   The $hd$ (head) and $tl$ (tail) operators are partial, and therefore raise the same problem as the predecessor function in the theory of natural numbers.

⟨ Derived operators and laws of sequences. 4.3.3.3 ⟩ ≡

$hd\,(\cdot)$   $: [\, s\,?\, sort \vdash [\, seq\,(s) \vdash s \,] \,]$
$;\, tl\,(\cdot)$   $: [\, s\,?\, sort \vdash [\, seq\,(s) \vdash seq(s) \,] \,]$
$;\, hd\_def : [\, s\,?\, sort\,;\, u\,?\, s\,;\, x\,?\, seq\,(s) \vdash hd(u :: x) = u \,]$
$;\, tl\_def : [\, s\,?\, sort\,;\, u\,?\, s\,;\, x\,?\, seq\,(s) \vdash tl(u :: x) = x \,]$

See also sections 4.3.3.4, 4.3.3.5, 4.3.3.6, and 4.3.3.7.

This code is used in section 4.3.3.

**4.3.3.4.** Another useful operator is the append function.

$\langle$ Derived operators and laws of sequences. 4.3.3.3 $\rangle+\equiv$
$; (\cdot) \frown (\cdot) : [\, s\,?\, sort \vdash [\, seq\,(s);\, seq(s) \vdash seq(s)\,]\,]$
$; def_\frown \qquad : [\, s\,?\, sort\,;\, l\,?\, seq\,(s)$
$\qquad\qquad \vdash\!\langle\; base := \langle\rangle \frown l = l$
$\qquad\qquad\quad , rec \;\; := [\, h\,?\, s\,;\, t\,?\, seq\,(s) \vdash (h :: t) \frown l = h :: (t \frown l)\,]$
$\qquad\qquad\quad \rangle$
$\qquad\qquad ]$

**4.3.3.5.** *len* counts the number of elements contained in a sequence.

$\langle$ Derived operators and laws of sequences. 4.3.3.3 $\rangle+\equiv$
$; len\,(\cdot) \;\; : [\, s\,?\, sort \vdash [\, seq\,(s) \vdash nat\,]\,]$
$; len\_def : \langle\; base := len\,\langle\rangle = 0, \quad rec := [\, s\,?\, sort\,;\, h\,?\, s\,;\, x\,?\, seq\,(s)$
$\qquad\qquad\qquad\qquad\qquad\qquad\qquad\qquad\qquad \vdash len\, h :: x = 1 + len\, x$
$\qquad\qquad\qquad\qquad\qquad\qquad\qquad\qquad\qquad ]$
$\qquad\qquad \rangle$

**4.3.3.6.** Similar to sets, one can define operators to filter a sequence by a predicate and to map a function over a sequence.

$\langle$ Derived operators and laws of sequences. 4.3.3.3 $\rangle+\equiv$
$; (\cdot) \rhd_{seq} (\cdot) : [\, s\,?\, sort \vdash [[s \vdash prop\,];\, seq(s) \vdash seq(s)\,]\,]$
$; (\cdot) *_{seq} (\cdot) \;\; : [s, t\,?\; sort \vdash [[s \vdash t];\, seq(s) \vdash seq(t)\,]\,]$
$; def_{\rhd_{seq}} \qquad : [\, s\,?\; sort\,;\, P\,?\, [s \vdash prop\,]$
$\qquad\qquad \vdash\!\langle\; base := P \rhd_{seq} \langle\rangle = \langle\rangle$
$\qquad\qquad\quad , rec \;\; := [\, h\,?\, s\,;\, t\,?\, seq\,(s)$
$\qquad\qquad\qquad\qquad \vdash\!\langle\; accept := [\, P\,(h) \vdash P \rhd_{seq} h :: t = h :: (P \rhd_{seq} t)\,]$
$\qquad\qquad\qquad\qquad\quad , reject \;\; := [\neg\; P\,(h) \vdash P \rhd_{seq} h :: t = P \rhd_{seq} t\,]$
$\qquad\qquad\qquad\qquad\quad \rangle$
$\qquad\qquad\qquad\qquad ]$
$\qquad\qquad\quad \rangle$
$\qquad\qquad ]$
$; def_{*_{seq}} \qquad : [s, t\,?\; sort\,;\, f\,?\, [s \vdash t]$
$\qquad\qquad \vdash\!\langle\; base := f *_{seq}\langle\rangle = \langle\rangle$
$\qquad\qquad\quad , rec \;\; := [\, u\,?\, s\,;\, x\,?\, seq\,(s) \vdash f *_{seq} u :: x = f(u) :: (f *_{seq} x)\,]$
$\qquad\qquad\quad \rangle$
$\qquad\qquad ]$

**4.3.3.7.** The function *elems* transforms a sequence into the set of its elements. The predicate *no_dupl* verifies that a sequence of sequences does not contain any duplicates (neither at the level of the sequences, nor at the level of the elements).

⟨ Derived operators and laws of sequences. 4.3.3.3 ⟩+ ≡

; *elems* : [ $s$ ? *sort* ⊢ [ $l$ : *seq* ($s$) ⊢ *set*($s$) ] ]

; *elems_def* :

⦇ *base* := *elems* (⟨⟩) = {}

, *rec* := [ $s$ ? *sort* ; $h$ ? $s$ ; $t$ ? *seq* ($s$) ⊢ *elems*($h$ :: $t$) = $h$ ⊙ *elems*($t$) ]

⦈

; *no_dupl* := [ $s$ ? *sort* ; $xx$ : *seq* (*seq*($s$))

⊢ *card* (*elems*($xx$)) = *len*($xx$) ∧

∀[ $x$ : *seq* ($s$) ⊢ $x$ ∈ *elems*($xx$) ⇒ *card*(*elems*($x$)) = *len*($x$) ]

]

### 4.3.4 Tuples

VDM offers a very general notion of composite object, including recursively defined data-structures. In this context we consider only the case of tuples of arbitrary but finite length. This will be sufficient for our examples.

Tuple types are formed from applying a tuple construction function to a list of sorts. For example, the text $T(nat \otimes list(nat) \otimes mt)$ denotes the type of all binary tuples with a natural number as first component and a list of naturals as second.

⟨ Tuples. 4.3.4 ⟩ ≡

**import context** *Tuples* :=

⟦ *sortlist* : **prim**

; *mt* : *sortlist*

; (·) ⊗ (·) : [*sort*; *sortlist* ⊢ *sortlist* ]

; $T$ : [*sortlist* ⊢ *sort* ]

; ⟨ Constructors of tuples. 4.3.4.1 ⟩

; ⟨ Axioms of tuples. 4.3.4.2 ⟩

; ⟨ Selection from a tuple. 4.3.4.3 ⟩

; ⟨ Derived properties of tuples. 4.3.4.4 ⟩

⟧

This code is used in section 4.1.2.

**4.3.4.1.** Actual tuples are formed from the two constructors *mkmt* (make empty tuple) and *mk* (make tuple). For example, the text $mk (0, mk(\langle\rangle, mkmt))$ denotes an element of the tuple type referred to above.

⟨ Constructors of tuples. 4.3.4.1 ⟩ ≡

  $mkmt$ : $T(mt)$

; $mk$     : [ $s$ ? $sort$ ; $sl$ ? $sortlist$ ⊢ [ $s$ ; $T(sl)$ ⊢ $T(s \otimes sl)$ ] ] ]

This code is used in section 4.3.4.

**4.3.4.2.**  As usual, the constructor $mk$ is required to be a one-to-one function.

⟨ Axioms of tuples. 4.3.4.2 ⟩ ≡

$mk\_injective$ : [ $s$ ? $sort$ ; $sl$ ? $sortlist$ ; $a, b$ ? $s$ ; $t_1, t_2$ ? $T(sl)$

$$\vdash \frac{mk\,(a, t_1) = mk(b, t_2)}{a = b \wedge t_1 = t_2}$$

]

This code is used in section 4.3.4.

**4.3.4.3.**  Finally, a selection function on tuples is specified.

⟨ Selection from a tuple. 4.3.4.3 ⟩ ≡

  $sel$     : [ $s$ ? $sort$ ; $sl$ ? $sortlist$

             ⊢⦇ $car$ := [ $T(s \otimes sl)$ ⊢ $s$ ], $cdr$ := [ $T(s \otimes sl)$ ⊢ $T(sl)$ ]⦈

             ]

; $sel\_def$ : [ $s$ ? $sort$ ; $sl$ ? $sortlist$ ; $a$ ? $s$ ; $t$ ? $T(sl)$

            ⊢⦇ $car$ := $sel$ . $car$ $(mk(a, t)) = a$

             , $cdr$ := $sel$ . $cdr$ $(mk(a, t)) = t$

             ⦈

            ]

This code is used in section 4.3.4.

**4.3.4.4.**  A simple tactic summaries the use of either of the defining equations of selection.

⟨ Derived properties of tuples. 4.3.4.4 ⟩ ≡

; $tuple\_unf\_tactic$ := **alt** [ $unfold(sel\_def.cdr)$, $unfold(sel\_def.car)$ ]

See also sections 4.3.4.5 and 4.3.4.6.

This code is used in section 4.3.4.

**4.3.4.5.** Finally, we define a special notation to construct binary tuples and state the corresponding injectivity property.

⟨ Derived properties of tuples. 4.3.4.4 ⟩+ ≡

$; (\cdot) \mapsto (\cdot) \qquad := [s, t ?\ sort \vdash [a : s ; b : t \vdash mk(a, mk(b, mkmt))]]$
$\qquad\qquad\qquad \therefore [s, t ?\ sort \vdash [s; t \vdash T(s \otimes t \otimes mt)]]$
$; mkbin\_injective : \quad [s, t ?\ sort ; a, c ?\ s ; b, d ?\ t$

$$\vdash \frac{(a \mapsto b) = (c \mapsto d)}{a = c \land b = d}$$

$]$

**4.3.4.6.** The following property relates binary tuples and set membership in a special situation.

⟨ Derived properties of tuples. 4.3.4.4 ⟩+ ≡

$; tuple\_prop : [s, t ?\ sort ; a ?\ s ; b ?\ t ; A ?\ set\,(T(s \otimes t \otimes mt))$

$$\vdash \frac{[c : T(s \otimes t \otimes mt) \vdash \dfrac{c \in A}{sel \,.\, car\,(sel.\,cdr\,(c)) = b}]}{[a \in sel.\,car * A \vDash (a \mapsto b) \in A]}$$

$]$

## 4.3.5  Finite Maps

Mappings in VDM are finite binary relations, i.e., finite sets of binary tuples, that satisfy the usual functional uniqueness constraint.

⟨ Finite maps. 4.3.5 ⟩ ≡
**import context** *FiniteMaps* :=
〚 *map* : [*sort*; *sort* ⊢ *sort*]
; ⟨ Constructors of finite maps. 4.3.5.1 ⟩
; ⟨ Axioms of finite maps. 4.3.5.2 ⟩
; ⟨ Derived operators of finite maps. 4.3.5.3 ⟩
〛

This code is used in section 4.1.2.

**4.3.5.1.** Finite maps are generated from the empty mapping *void* and the insertion ⊕ of a new pair into a map.

⟨Constructors of finite maps. 4.3.5.1⟩ ≡

$void \quad : [s, t\,?\ sort \vdash map(s, t)]$

$; (\cdot) \oplus (\cdot) : [s, t\,?\ sort \vdash [T(s \otimes t \otimes mt); map(s, t) \vdash map(s, t)]]$

$; dom \quad : [s, t\,?\ sort \vdash [map\,(s, t) \vdash set(s)]]$

This code is used in section 4.3.5.

**4.3.5.2.** The absorption and commutativity properties are slightly different from those for sets. The (nonstandard) induction rule reflects the absorption law by an additional hypothesis using the function to compute the domain of a mapping.

⟨Axioms of finite maps. 4.3.5.2⟩ ≡

$map\_overwrite : [s, t\,?\ sort;\ x\,?\ s;\ y, z\,?\ t;\ m\,?\ map\,(s, t)$
$$\vdash (x \mapsto z) \oplus (x \mapsto y) \oplus m = (x \mapsto y) \oplus m$$
$]$

$; map\_commute : [s, t\,?\ sort;\ x_1, x_2\,?\ s;\ y, z\,?\ t;\ m\,?\ map\,(s, t)$
$$\vdash \dfrac{\neg\,x_1 = x_2}{(x_1 \mapsto z) \oplus (x_2 \mapsto y) \oplus m = (x_2 \mapsto y) \oplus (x_1 \mapsto z) \oplus m}$$
$]$

$; map\_induction : [s, t\,?\ sort;\ P\,?\ [map\,(s, t) \vdash prop]$
$$\vdash \dfrac{P\,(void);\ [\,x : s;\,y : t;\,m : map\,(s, t) \vdash \dfrac{P\,(m);}{\neg\,x \in dom(m)}\ ]}{[\,m\,?\ map\,(s, t) \vdash P(m)]}$$
$]$

$; map\_dom \quad : \langle\!\langle base := dom\,(void) = \{\}$
$\qquad , rec := [s, t\,?\ sort;\ x\,?\ s;\ y\,?\ t;\ m\,?\ map\,(s, t)$
$\qquad\qquad \vdash dom\,((x \mapsto y) \oplus m) = x \odot dom(m)$
$\qquad\qquad ]$
$\qquad \rangle\!\rangle$

This code is used in section 4.3.5.

**4.3.5.3.** The application of a mapping to an argument is recursively defined as follows.

⟨Derived operators of finite maps. 4.3.5.3⟩ ≡

$apply \quad : [s, t\,?\ sort \vdash [map\,(s, t); s \vdash t]]$

; $apply\_def$ : $[s, t\, ?\ sort\, ;\ x\, ?\ s\, ;\ y\, ?\ t\, ;\ m\, ?\ map\,(s, t)$

$$\vdash\!\langle\ first\ :=\ \left|\frac{\neg\ x \in dom(m)}{apply\,((x \mapsto y) \oplus m, x) = y}\right.$$

$$,\ rec\ :=[\,z\,?\,s\ \vdash\ \left|\frac{\neg\ x = z}{apply\,((z \mapsto y) \oplus m, x) = apply\,(m, x)}\right.]$$

$$\flat$$

$$]$$

See also section 4.3.5.4.

This code is used in section 4.3.5.

**4.3.5.4.** Several other operators are defined: $rng$ computes the range of a mapping, $\lhd$ restricts a mapping to a given domain, and $\lhd\!\!\!-$ deletes pairs whose left-hand value belongs to a given domain.

⟨ Derived operators of finite maps. 4.3.5.3 ⟩+ ≡

; $rng$ $\qquad\quad := [s, t\,?\ sort \vdash [\,m : map\,(s, t) \vdash apply\,(m) * dom(m)\,]\,]$
; $(\cdot) \lhd (\cdot)$ $\qquad : [s, t\,?\ sort \vdash [\,set\,(s); map(s, t) \vdash map(s, t)\,]\,]$
; $dom\_restriction$ :
$[s, t\,?\ sort\,;\ A\,?\ set\,(s)$
$\vdash\!\langle\ base := A \lhd void = void$

$\quad,\ rec\quad := [\,x\,?\ s\,;\ y\,?\ t\,;\ m\,?\ map\,(s, t)$

$$\vdash\!\langle\ accept\ :=\ \left|\frac{x \in A}{A \lhd ((x \mapsto y) \oplus m) = (x \mapsto y) \oplus (A \lhd m)}\right.$$

$$,\ reject\ :=$$

$$\left|\frac{\neg\ x \in A}{A \lhd ((x \mapsto y) \oplus m) = (A \lhd m)}\right.$$

$$\flat$$

$$]$$

$$\flat$$

$$]$$

; $(\cdot) \lhd\!\!\!- (\cdot)$ $\qquad\quad := [s, t\,?\ sort$
$$\qquad\qquad\qquad \vdash [\,A : set\,(s); m : map\,(s, t) \vdash (dom(m) \setminus A) \lhd m\,]$$
$$\qquad\qquad\qquad ]$$

## 4.3.6  Simple Tactics

Finally, some tactics involving the VDM datastructures are summarized. These tactics are not very sophisticated, however they are structured and they give some idea of how a large and sophisticated library of tactics could look like. The material has been composed based on the needs of the VDM case study presented in Chap. 5.

Two simple but useful tactics describe the decomposition of equations and the evaluation of VDM operations. A third tactic combines these two with the tactic for logical simplifications. This tactic will be used for the proof demonstrations in Sect. 5.5.7. Warning: The rules in the second and third tactic are laid out in a two-dimensional style, do not read them line-by-line but column-by-column.

⟨ VDM tactics. 4.3.6 ⟩ ≡
**context** *VDMTactics* :=
⟦ *VDMDecomposition* :=
  **alt** [*psubst*(*peano₄*), *psubst*(*mk_injective*), *psubst*(*seq_free₂*)]
; *VDMEvaluation*      :=
  ⟨ *base* := **alt** [*rsubst*(*card_def*.*base*) ,    *rec* := **alt** [*rsubst*(*card_def*.*rec*)
              , *rsubst*(*def₊*.*base*)                      , *rsubst*(*def₊*.*rec*)
              , *rsubst*(*def▷*.*base*)                      , *rsubst*(*def₊_seq*.*rec*)
              , *rsubst*(*def₊_seq*.*base*)                  , *rsubst*(*map_dom*.*rec*)
              , *rsubst*(*def▷_seq*.*base*)                  ]
              , *rsubst*(*map_dom*.*base*)
              ]
  ⟩
; *VDMSimplification*  :=
  ⟨ *base* := **alt** [*LogicalSimplification*  ,    *rec* := **alt** [*LogicalSimplification*
              , *VDMDecomposition*                        , *VDMDecomposition*
              , *VDMEvaluation*. *base*                   , *VDMEvaluation*. *rec*
              ]                                           ]
  ⟩
⟧

This code is used in section 4.1.2.

## 4.4   Basic Theories for Algorithm Calculation

**4.4.1.**    ⟨ Imports needed by *CalculationalBasics*. 4.4.1 ⟩ ≡
**import** *LogicalBasis*

This code is used in section 4.1.3.

### 4.4.1 Extensional Equality of Terms or Functions

In order to have easy access to both notions of equality in the case study on algorithm calculation, we combine the two into one. To do so, we first import the contexts defined in Secs. 4.2.2 and 4.2.3 and rename the declared variables so that name clashes are avoided.

**4.4.1.1.** 〈Extensional equality of terms or functions. 4.4.1.1〉 ≡
**import context** *ExtensionalEquality* :=
[ **import** *EqualityOfTerms* [= =:*teq*, *refl* =: *trefl*,
$\qquad\qquad\qquad$ *subst* =:*tsubst*, *sym* =: *tsym*,
$\qquad\qquad\qquad$ *trans* =:*ttrans*, *unfold* =: *tunfold*,
$\qquad\qquad\qquad$ *rsubst* =:*trsubst*, *fold* =: *tfold* ]
; **import** *EqualityOfFunctions* [= =:*feq*, *refl* =: *frefl*,
$\qquad\qquad\qquad$ *subst* =:*fsubst*, *sym* =: *fsym*,
$\qquad\qquad\qquad$ *trans* =:*ftrans*, *unfold* =: *funfold*,
$\qquad\qquad\qquad$ *rsubst* =:*frsubst*, *fold* =: *ffold* ]
;〈Principle of extensionality. 4.4.1.2〉
;〈Overloading equality. 4.4.1.3〉
;〈Unfold rules for the overloaded equality. 4.4.1.4〉
;〈Fold rules for the overloaded equality. 4.4.1.5〉
]

This code is used in section 4.1.3.

**4.4.1.2.** The principle of *extensionality* relates term equality and functional equality. It states that two functions are considered to be equal if they return the same values on their domain of definition.

〈Principle of extensionality. 4.4.1.2〉 ≡

$$extensionality \;:[s,t\;?\;\;sort\;;f,g\;?\;[s\vdash t]\vdash \frac{[\,x\;?\;s\;\vdash f(x)\;teq\;g(x)\,]}{f\;feq\;g}]$$

This code is used in section 4.4.1.1.

**4.4.1.3.** It is rather tedious to always distinguish explicitly between the two equalities, in particular since the equality in question can always be deduced from the context by looking at the type of the left- or the right-hand side. This is a typical situation which calls for the use of Deva's alternative construct to leave the task of determining the concrete equality up to the system.

〈Overloading equality. 4.4.1.3〉 ≡

$(\cdot) = (\cdot) := $ **alt** $[teq, feq]$
$; \ subst \quad := $ **alt** $[tsubst, fsubst]$
$; \ sym \qquad := $ **alt** $[tsym, fsym]$
$; \ trans \quad := $ **alt** $[ttrans, ftrans]$
$; \ rsubst \quad := $ **alt** $[trsubst, frsubst]$

This code is used in section 4.4.1.1.

**4.4.1.4.** Four different unfold rules can be defined for this new equality because the two equalities needed for the unfold rule may both be either equalities between terms or equalities between functions. Two rules are already defined, namely *tunfold* and *funfold*. The remaining two can be derived by a suitable instantiation of the two substitution principles. Combining all four specific unfold rules gives a general unfold rule.

$\langle$ Unfold rules for the overloaded equality. 4.4.1.4 $\rangle \equiv$

   *tfunfold* :=

   $[s, t, u ? \ sort ; \ h \ ? \ [t \vdash u]; \ F \ ? \ [s \vdash [t \vdash u]]$
   $\vdash subst \ (P := [\, z : s \ \vdash h = F(z)\,])$
   $]$

      $\therefore [s, t, u ? \ sort ; x, y ? \ s ; h ? [t \vdash u]; F ? [s \vdash [t \vdash u]] \vdash [x = y \vdash \begin{vmatrix} h = F(x) \\ h = F(y) \end{vmatrix}]]$

   $; \ ftunfold :=$

   $[s, t, u ? \ sort ; \ z \ ? \ u; \ F \ ? \ [[s \vdash t] \vdash u]$
   $\vdash subst \ (P := [\, h : [s \vdash t] \vdash z = F(h)\,])$
   $]$

      $\therefore [s, t, u ? \ sort ; f, g ? [s \vdash t]; z ? \ u ; F ? [[s \vdash t] \vdash u] \vdash [f = g \vdash \begin{vmatrix} z = F(f) \\ z = F(g) \end{vmatrix}]]$

   $; \ unfold \quad := $ **alt** $[tunfold, tfunfold, ftunfold, funfold]$

This code is used in section 4.4.1.1.

**4.4.1.5.** In a similar manner, a generalized fold rule is constructed.

$\langle$ Fold rules for the overloaded equality. 4.4.1.5 $\rangle \equiv$

   *tffold* := $tsym \gg tfunfold$

      $\therefore [s, t, u ? \ sort ; x, y ? \ s ; \ h \ ? [t \vdash u]; \ F \ ? [s \vdash [t \vdash u]]$

        $\vdash [x = y \vdash \begin{vmatrix} h = F(y) \\ h = F(x) \end{vmatrix}]$

      $]$

; *ftfold* := *fsym* ⬦ *ftunfold*

$$\therefore [s, t, u ? \; sort ; f, g ? [s \vdash t]; \; z \; ? \; u ; \; F \; ? [[s \vdash t] \vdash u]$$

$$\vdash [f = g \vdash \left| \begin{array}{l} z = F(g) \\ \hline z = F(f) \end{array} \right| ]$$

]

; *fold*  := **alt** [*tfold, tffold, ftfold, ffold*]

This code is used in section 4.4.1.1.

The rules *subst, rsubst, unfold,* and *fold* suffice to perform all the equality reasoning steps in Chap. 6.

## 4.4.2  Terms Involving Functions

The most straightforward way to formalize functions in Deva is to identify them with abstractions, i.e., a function from sort *s* to sort *t* is of type [*s* ⊢ *t*], and this is exactly the way we have treated them so far. Of course, we now have to externalize a few other concepts such as the identity function or functional composition, but this is easily done. A notion which is related to functions and which we will need in Chap. 6 is that of injectivity. Extensionality can be used to derive that *id* is both a left and a right identity of composition.

**4.4.2.1.**  ⟨ Terms involving functions. 4.4.2.1 ⟩ ≡
**import context** *FunctionalTerms* :=
[ *id*       := [ *s* ? *sort* ⊢ [ *x* : *s* ⊢ *x* ]]
; (·) ∘ (·)  := [*s, t, u* ? *sort* ; *f* : [*t* ⊢ *u*]; *g* : [*s* ⊢ *t*] ⊢ [ *x* : *s* ⊢ *f(g(x))*]]]

; *injective* := [*s, t* ? *sort* ; *f* : [*s* ⊢ *t*]; *x, y* ? *s* ⊢ $\left| \begin{array}{l} f(x) = f(y) \\ \hline x = y \end{array} \right|$ ]

; *pid*      := [*s, t* ? *sort* ; *f* ? [*s* ⊢ *t*]
        ⊢ ⟨*left* := [ *x* ? *s* ⊢ *trefl* ∴ (*id* ∘ *f*)(*x*) = *f(x)* ]
            \ *extensionality. down*
               ∴ *id* ∘ *f* = *f*
           , *right* := [ *x* ? *s* ⊢ *trefl* ∴ (*f* ∘ *id*)(*x*) = *f(x)* ]
            \ *extensionality. down*
               ∴ *f* ∘ *id* = *f*

        ⟩
        ]

]

This code is used in section 4.1.3.

Now, the advantage of this formalization becomes apparent because the following laws are for free due to the conversion laws of Deva:

$$f \circ id \simeq f,$$
$$id \circ f \simeq f,$$
$$f \circ (g \circ h) \simeq (f \circ g) \circ h.$$

These "automatic" laws will significantly reduce the size of the formal calculations.

### 4.4.3  Some Bits of Algebra

**4.4.3.1.**     ⟨ Some Bits of Algebra. 4.4.3.1 ⟩ ≡
**import context** *Algebra* :=
[[ ⟨ Generic algebraic properties. 4.4.3.2 ⟩
; ⟨ Monoid. 4.4.3.3 ⟩
; ⟨ Distributive monoid pair. 4.4.3.4 ⟩
; ⟨ Boolean algebra. 4.4.3.5 ⟩
]]

This code is used in section 4.1.3.

**4.4.3.2.**   In this section we formalize a few primitive algebraic notions concerning binary operators. First, we characterize units and zeros of binary operators, associativity, commutativity, absorption, distributivity, and finally a generalized form of distributivity.

⟨ Generic algebraic properties. 4.4.3.2 ⟩ ≡

$$
\begin{aligned}
zero \qquad &:= [\, s \,?\, sort \,;\, (\cdot) \otimes (\cdot) : [s; s \vdash s]; 0_\otimes : s \\
&\quad \vdash [\, a \,?\, s \vdash \langle\!| left := (0_\otimes \otimes a) = 0_\otimes, right := (a \otimes 0_\otimes) = 0_\otimes |\!\rangle\,] \\
&\quad ] \\
;\ unit \qquad &:= [\, s \,?\, sort \,;\, (\cdot) \otimes (\cdot) : [s; s \vdash s]; 1_\otimes : s \\
&\quad \vdash [\, a \,?\, s \vdash \langle\!| left := (1_\otimes \otimes a) = a, right := (a \otimes 1_\otimes) = a |\!\rangle\,] \\
&\quad ] \\
;\ associative \quad &:= [\, s \,?\, sort \,;\, (\cdot) \oplus (\cdot) : [s; s \vdash s] \\
&\quad \vdash [a, b, c \,?\, s \vdash ((a \oplus b) \oplus c) = a \oplus (b \oplus c)] \\
&\quad ] \\
;\ commutative \quad &:= [\, s \,?\, sort \,;\, (\cdot) \oplus (\cdot) : [s; s \vdash s] \\
&\quad \vdash [a, b \,?\, s \vdash (a \oplus b) = b \oplus a] \\
&\quad ]
\end{aligned}
$$

```
; absorptive      := [ s ? sort ; (·) ⊕ (·), (·) ⊗ (·) : [s; s ⊢ s]
                      ⊢[a, b ? s ⊢ (a ⊕ (a ⊗ b)) = a]
                     ]
; distributive    := [ s ? sort ; (·) ⊗ (·), (·) ⊕ (·) : [s; s ⊢ s]
                      ⊢[a, b, c ? s ⊢ (a ⊗ (b ⊕ c)) = (a ⊗ b) ⊕ (a ⊗ c)]
                     ]
; gen_distributive := [ s ? sort ; f : [s ⊢ s]; (·) ⊕ (·), (·) ⊗ (·) : [s; s ⊢ s]
                      ⊢[a, b ? s ⊢ (f(a ⊕ b)) = f(a) ⊗ f(b)]
                     ]
```

This code is used in section 4.4.3.1.

**4.4.3.3.**  A monoid is given by an associative binary operator which has a unit.

```
⟨ Monoid. 4.4.3.3 ⟩ ≡
monoid := [ s ? sort ; ⊕ : [s; s ⊢ s]; 1⊕ : s
            ⊢⦇ unit := unit (⊕, 1⊕), associative := associative (⊕)⦈
           ]
```

This code is used in section 4.4.3.1.

**4.4.3.4.**  Two monoids form a *distributive monoid pair* if one operator distributes over the other.

```
⟨ Distributive monoid pair. 4.4.3.4 ⟩ ≡
distrib_monoids :=
[ s ? sort ; ⊕ : [s; s ⊢ s]; 1⊕ : s ; ⊗ : [s; s ⊢ s]; 1⊗ : s
⊢⦇ mon_plus   := monoid (⊕, 1⊕)   , mon_times := monoid (⊗, 1⊗)
  , left_distrib := distributive (⊗, ⊕)
  ⦈
]
```

This code is used in section 4.4.3.1.

**4.4.3.5.**  Similarly, we combine all the properties needed to specify a boolean algebra.

```
⟨ Boolean algebra. 4.4.3.5 ⟩ ≡
boolean_algebra :=
```

```
[ s ? sort ; ⊥, ⊤ : s ; ⊔, ⊓ : [s; s ⊢ s];  − : [s ⊢ s]
⊢⟨ unit      := ⟨ bot := unit (⊔, ⊥)   , assoc  := ⟨ join := associative (⊔)
             , top := unit (⊓, ⊤)                 , meet := associative (⊓)
             ⟩                                     ⟩
  , zero     := ⟨ bot := zero (⊓, ⊥)   , commut := ⟨ join := commutative (⊔)
             , top := zero (⊔, ⊤)                 , meet := commutative (⊓)
             ⟩                                     ⟩
  , idem     := [ a ? s ⊢ −(−(a)) = a ], absorp  := ⟨ join := absorptive (⊔, ⊓)
                                                   , meet := absorptive (⊓, ⊔)
                                                   ⟩
  , distrib  := ⟨ join := distributive (⊔, ⊓)
             , meet := distributive (⊓, ⊔)
             ⟩
  , morgan   := ⟨ join := gen_distributive (−, ⊔, ⊓)
             , meet := gen_distributive (−, ⊓, ⊔)
             ⟩
  ⟩
]
```

See also sections 4.4.3.6, 4.4.3.7, and 4.4.3.8.

This code is used in section 4.4.3.1.

**4.4.3.6.**    The properties defined in the previous section can be used to perform simplifications within boolean expressions containing the minimal element ⊥ or the maximal element ⊤. This is suitably done by defining a tactic which combines the appropriate rules. Note that, according to the type of the *unfold* law (Sect. 4.2.2), the transformation is restricted to the right-hand side of equations, i.e., the law $x = y$ is used to derive $z = F(y)$ from $z = F(x)$. This tactic will be used several times in the examples of Chap. 6.

⟨ Boolean algebra. 4.4.3.5 ⟩+ ≡

```
; bool_simp :=
[ s ? sort ; ⊥, ⊤ ? s ; ⊔, ⊓ ? [s; s ⊢ s];  − ? [s ⊢ s]
⊢[ ba : boolean_algebra (⊥, ⊤, ⊔, ⊓, −)
  ⊢ alt [unfold(ba. unit . bot .left) , unfold(ba. unit . bot .right)
       , unfold(ba. unit . top .left), unfold(ba. unit . top .right)
       , unfold(ba. zero . bot .left), unfold(ba. zero . bot .right)
       , unfold(ba. zero . top .left), unfold(ba. zero . top .right)
       ]
  ]
]
```

**4.4.3.7.** Another useful transformation tactic moves negations inward, using the de-Morgan law, and removes double negations as often as possible.

⟨Boolean algebra. 4.4.3.5 ⟩+ ≡
; *bool_neg_simp* :=
  [ *s* ? *sort* ; ⊥, ⊤ : *s* ; ⊔, ⊓ ? [*s*; *s* ⊢ *s*]; − ? [*s* ⊢ *s*]
  ⊢[ *ba* : *boolean_algebra* (⊥, ⊤, ⊔, ⊓, −)
    ⊢ **alt** [*unfold*(*ba*. *morgan* .*join*), *unfold*(*ba*. *morgan* .*meet*)
          , *unfold*(*ba*.*idem*)
          ]
  ]
]

**4.4.3.8.** Finally, there is a tactic which summarizes the application of associativity and commutativity.

⟨Boolean algebra. 4.4.3.5 ⟩+ ≡
; *bool_ac* :=
  [ *s* ? *sort* ; ⊥, ⊤ ? *s* ; ⊔, ⊓ ? [*s*; *s* ⊢ *s*]; − ? [*s* ⊢ *s*]
  ⊢[ *ba* : *boolean_algebra* (⊥, ⊤, ⊔, ⊓, −)
    ⊢ **alt** [*unfold*(*ba*. *assoc* .*join*)    , *fold*(*ba*. *assoc* .*join*)
          , *unfold*(*ba*. *assoc* .*meet*)   , *fold*(*ba*. *assoc* .*meet*)
          , *unfold*(*ba*. *commut* .*join*), *unfold*(*ba*. *commut* .*meet*)
          ]
  ]
]

## 4.4.4   Induced Partial Ordering

In this section we want to formalize the partial order which is induced by a boolean algebra. This partial order is characterized by the equivalence

$$a \sqsubseteq b \quad \equiv \quad a \sqcap b = a \quad \equiv \quad a \sqcup b = b.$$

⟨Induced Partial Ordering. 4.4.4 ⟩ ≡
**import context** *PartialOrdering* :=
[ *s*                      ?   *sort*
; ⊥, ⊤                     ?   *s*
; (·) ⊔ (·), (·) ⊓ (·) ?   [*s*; *s* ⊢ *s*]
; (·) − (·)                ?   [*s* ⊢ *s*]

; $ba$                :    $boolean\_algebra\ (\bot, \top, \sqcup, \sqcap, -)$
; $(\cdot) \sqsubseteq (\cdot)$       $:= [\,a, b : s \vdash a \sqcup b = b\,]$
; $\langle$ Some laws of partial orderings. 4.4.4.1 $\rangle$
$]\!]$

This code is used in section 4.1.3.

We declared the first six variables implicitly in coordination with the type of *boolean_algebra*. We are now able to import this context and instantiate *ba* with any boolean algebra, obtaining access to the following laws.

**4.4.4.1.** For example, reflexivity and transitivity can be proven without problems.

$\langle$ Some laws of partial orderings. 4.4.4.1 $\rangle \equiv$

$refl\_smth \quad := [\,a\ ?\ s$
          $\vdash trefl$
              $\therefore a = a$
          $\backslash\ fold(ba.\ absorp\ .join)$
              $\therefore a = a \sqcup (a \sqcap \top)$
          $\backslash\ unfold(ba.\ unit\ .\ top\ .right)$
              $\therefore a = a \sqcup a$
          $]$
            $\therefore [\,a\ ?\ s \vdash a \sqsubseteq a\,]$
; $trans\_smth := [\,a, b, c\ ?\ s\ ;\ hyp_1 : a \sqsubseteq b;\ hyp_2\ : b \sqsubseteq c$
          $\vdash trefl$
            $\therefore (a \sqcup c) = a \sqcup c$
          $\backslash\ fold(hyp_2)$
            $\therefore (a \sqcup c) = a \sqcup (b \sqcup c)$
          $\backslash\ fold(ba.\ assoc\ .join)$
            $\therefore (a \sqcup c) = (a \sqcup b) \sqcup c$
          $\backslash\ unfold(hyp_1)$
            $\therefore (a \sqcup c) = b \sqcup c$
          $\backslash\ unfold(hyp_2)$
            $\therefore (a \sqcup c) = c$
      $]$
            $\therefore [\,a, b, c\ ?\ s \vdash \dfrac{a \sqsubseteq b;\ b \sqsubseteq c}{a \sqsubseteq c}\,]$

See also section 4.4.4.2.

This code is used in section 4.4.4.

**4.4.4.2.** This section concludes by stating two properties which are needed later on. The first states that $\bot$ is at most $\top$; the second, that under a certain condition, meet and join may be exchanged. A proof for the latter property is not given but is left as a simple exercise for the reader.

$\langle$ Some laws of partial orderings. 4.4.4.1 $\rangle + \equiv$

; *triv* $:= ba \, . \, zero \, . \, top \, . right$

$$\therefore \bot \sqsubseteq \top$$

; *exch_meet_join* $: \quad [a, b, c \, ? \, s \vdash \dfrac{a \sqsubseteq b}{(b \sqcap (a \sqcup c)) = a \sqcup (b \sqcap c)}]$

# 5 Case Study on VDM-Style Developments

The development and application of systematic methods for the construction of safe software systems is a wide and active area of research in computer science. In particular, the VDM methodology [60] has been developed up to a point where it begins to enter industrial application. VDM is a "rigorous" methodology that defines systematic transitions from abstract formal specifications of software systems to concrete implementations. This chapter presents a case study on the complete formalization of VDM-style developments. VDM was chosen as topic of a case study because it has been developed and taught for many years. Nowadays, it has reached a reasonable level of maturity and acceptance. Moreover, VDM is a good target method for complete formalization since it precisely and systematically describes how development proof obligations are constructed.

In a first phase of this case study, VDM proof obligations were developed using Deva, but the theorems to be proved were determined *outside* the framework [66]. Later, VDM was described formally itself, and the proof obligations were fully integrated into the framework. As a consequence of this formalization, it becomes possible to formally reason *about* VDM, e.g. to derive fundamental properties. As an illustration, this chapter reports on an exercise to study how a typical property of data-refinement techniques, namely transitivity of data-refinement, is formally verified to be a property of data-refinement in VDM as formalized in the present context.

Concurrently to the formalization activity, an interesting biological case study, the *Human Leukocyte Antigen (HLA) typing problem*, was developed in VDM [71]. It can be sketched as follows: "Given the biological rules managing the reactions between sera and cells, one version of the HLA typing problem is the determination of the antibodies belonging to a serum. This determination is worked out from the results of biological experiments which consists of observing the reaction of the serum with a set of cells". This development example will also be presented, as an illustration of the VDM formalization.

In its early phase this case study was based on the experience gained from using the B tool to express and reuse VDM developments [77]. An earlier version of the Deva formalization and a B formalization of VDM are compared in [68].

## 5.1 Overview

The chapter is organized as follows: After a short introduction to the Vienna Development Method (VDM), a formalization of reification in VDM is presented in Deva. Then, the HLA typing case study is introduced and a VDM development for this case study is sketched in a semi-formal way. After that, at the heart of this chapter, a formalized development in Deva is presented. Finally, a proof is given in Deva that VDM-reification is transitive. The overall structure of the Deva formalization looks as follows:

**context** *VDMCaseStudy* :=
⟦ **import** *BasicTheories*

; **import**   *VDMBasics*
; **import**   *VDMTactics*
; ⟨ VDM reification. 5.3 ⟩
; ⟨ Human leukocyte antigen typing problem. 5.5 ⟩
; ⟨ Transitivity of reification. 5.6.3.2 ⟩
]

## 5.2   The Vienna Development Method

This section presents a short reminder of the Vienna Development Method. A complete description may be found in [59].

VDM is denotational and model-based. Operations are described over a *state* modeling the system to be described. The state is a composite data-type, possibly restricted by a predicate called the *invariant*. Each operation is described by a pair of predicates, the *precondition* and the *postcondition*. The precondition describes the states in which the operation can occur, and the postcondition describes the relationship between the original and resulting states. For an operation to be valid, certain conditions, the *proof obligations*, must be met.

In VDM, *data reification* supports the refinement of data structures: starting with the original *abstract description of data*, certain design decisions are made and a more *concrete description* is produced. Next, a *retrieve function* which relates the concrete state to the abstract state is given. The concrete description can then be shown to *satisfy* the abstract description by discharging *proof obligations* for each operation. The concrete description becomes the abstract description for the next step in the development, and the process is continued until a satisfactory implementation is reached. *Operation decomposition*, which allows the refinement of control structures, is also provided in VDM, however not tackled in this case study.

## 5.3   Formalization of VDM-Reification in Deva

The general structure of the formalization was explained in the introduction. The present section focuses on the formalization of the reification aspect of VDM. The definition of the VDM data reification itself consists of defining VDM operations, VDM versions, and finally the notion of reification between two versions:

⟨ VDM reification. 5.3 ⟩ ≡
**import context** *VDMReification* :=
[[ ⟨ VDM operations. 5.3.1.4 ⟩
; ⟨ VDM versions. 5.3.2.4 ⟩
; ⟨ VDM version reification. 5.3.3.6 ⟩
]

This code is used in section 5.1.

### 5.3.1 Operations

**5.3.1.1.** Operations in VDM are described as state transformers: An operation transforms a given input state and input parameter into an output state and output parameter. A precondition is imposed on the input parameter and the input state, and a postcondition relates the given input parameter and input state to an output parameter and output state. The signature of VDM-operations can be described in Deva as follows:

$\langle$ Signature of VDM operations. 5.3.1.1 $\rangle \equiv$

$$OP := [in, out, state : sort \vdash \frac{in; state}{\langle\!\langle pre := prop, post := [out; state \vdash prop]\rangle\!\rangle}]$$

This code is used in section 5.3.1.4.

For an operation with a given state sort $state$ and parameter sorts $in$ and $out$, the text $OP\ (in, out, state)$ denotes the signatures of the pre- and postcondition of that operation.

**5.3.1.2.** VDM-operations may have no parameters at all. A dummy sort $void$ is introduced to be used as parameter sort in such a case.

$\langle$ Empty sort. 5.3.1.2 $\rangle \equiv$

$void\ :\ sort$

This code is used in section 5.3.1.4.

**5.3.1.3.** In VDM, operations do not work on all states, but only on those satisfying an invariant. For any operation $op$, two conditions involving invariants, pre- and postconditions are defined:

- For a given input of $op$ on which the precondition holds and whose state component satisfies the invariant, there exists an output of $op$ which is related to the input by the postcondition (*satisfiability*).
- For given input and output on which pre- and postcondition hold, the operation $op$ preserves the invariants (*preservation*).

These conditions can be formally stated as follows:

$\langle$ Proof obligations for operations. 5.3.1.3 $\rangle \equiv$

$op\_valid :=$

$[\,in, out, state\,?\,sort\,;\,op\,:\,OP\,(in, out, state);\,inv\,:[state \vdash prop\,]$
$\vdash[\,i\,:\,in\,;\,st_i\,:\,state$

$\qquad inv\,(st_i);\,op(i, st_i).\,pre$

$\qquad \langle\!|\,satisfiability\,:= \exists_2[\,o\,:\,out\,;\,st_o\,:\,state \vdash op(i, st_i).\,post\,(o, st_o)\,]$

$$\vdash \quad ,\,preservation\,:= [\,o\,:\,out\,;\,st_o\,:\,state \vdash \frac{op\,(i, st_i).\,post\,(o, st_o)}{inv\,(st_o)}]$$

$\qquad |\!\rangle$

$\qquad ]$

$]$

This code is used in section 5.3.1.4.

**5.3.1.4.**   Finally, the given pieces are assembled into a context formalizing operations.

$\langle$ VDM operations. 5.3.1.4 $\rangle \equiv$
$[\![\,\langle$ Signature of VDM operations. 5.3.1.1 $\rangle$
$;\langle$ Empty sort. 5.3.1.2 $\rangle$
$;\langle$ Proof obligations for operations. 5.3.1.3 $\rangle$
$]\!]$

This code is used in section 5.3.

### 5.3.2   Versions

**5.3.2.1.**   A *version* is a list $op_1 \odot (op_2 \odot \ldots (op_n \odot [\,]) \ldots)$ of operations together with an invariant *inv*. The invariant can be seen as a parameter of the list. The type of versions with invariant *inv*, denoted by *version (inv)*, together with the constructors $[\,]$ and $\odot$ to build versions can be introduced in Deva as follows:

$\langle$ Construction of versions. 5.3.2.1 $\rangle \equiv$
$[\![\,version\,:[\,state\,?\,sort \vdash [[state \vdash prop\,] \vdash \mathbf{prim}\,]\,]$
$;[\,]\qquad\quad :[\,state\,?\,sort\,;\,inv\,?\,[state \vdash prop\,] \vdash version(inv)\,]$
$;\,(\cdot)\odot(\cdot)\,:[in, out, state\,?\,sort\,;\,inv\,?\,[state \vdash prop\,]$
$\qquad\qquad\qquad \vdash[\,OP\,(in, out, state);\,version(inv) \vdash version(inv)\,]$
$\qquad\qquad\qquad ]$
$]\!]$

This code is used in section 5.3.2.4.

**5.3.2.2.** A version $v$ of type *version* $(inv)$ is *valid*, denoted by $v \checkmark$, if and only if all its operators $op$ satisfy the operator condition *op_valid* $(op, inv)$ for operations. $v \checkmark$ can be specified as follows:

⟨ Proof obligations for versions. 5.3.2.2 ⟩ ≡

$[\![ (\cdot) \checkmark \ : [\, state \ ? \ sort \, ; \ inv \ ? \, [\, state \vdash prop \,] $
$\qquad \vdash [\, version \, (inv) \vdash prop \,] $
$\qquad ] $
$; \ def_\checkmark \ : \{\! empty := [\,] \checkmark $
$\qquad , \ cons \quad := [\, in, out, state \, ? \ sort \, ; \qquad\qquad op \ ? \ OP \, (in, out, state) $
$\qquad\qquad ; \ inv \qquad\quad ? \, [\, state \vdash prop \,]; \ v \quad ? \ version \, (inv) $
$\qquad\qquad \vdash \dfrac{\Big| \, (op \odot v) \checkmark}{\{\! base := v \checkmark \, , newop := op\_valid \, (op, inv)\!\}} $
$\qquad\qquad ] $

$\qquad \}$

$]\!]$

This code is used in section 5.3.2.4.

The reader might wonder, why not simply view versions as sequences and reuse the notations and axioms of the data type *Sequences* defined in Sect. 4.3.3. Unfortunately, while *Sequences* is about lists that are parametric over *sorts*, our formalization models versions as parametric over *abstractions from sorts to propositions*. Therefore, we were forced to introduce new constructors to build versions.

**5.3.2.3.** A principle of structural induction on versions can be defined as follows:

⟨ Induction on versions. 5.3.2.3 ⟩ ≡

*version_induction* :

$[\, state \ ? \ sort \, ; \ inv \ ? \, [\, state \vdash prop \,]; \ P \ ? \, [\, version \, (inv) \vdash prop \,] $
$\qquad \Big| \ P \, ([\,]); $
$\vdash \Big[ in, out \, ? \ sort \, ; v : version \, (inv); op : OP \, (in, out, state) \vdash \dfrac{P \, (v)}{P \, (op \odot v)} \Big] $
$\qquad\qquad\qquad\qquad [\, v \ ? \ version \, (inv) \vdash P(v) \,] $
$] $

This code is used in section 5.3.2.4.

**5.3.2.4.** Finally, the given pieces are assembled into a context formalizing versions in VDM.

⟨ VDM versions. 5.3.2.4 ⟩ ≡
⟦ ⟨ Construction of versions. 5.3.2.1 ⟩
; ⟨ Proof obligations for versions. 5.3.2.2 ⟩
; ⟨ Induction on versions. 5.3.2.3 ⟩
⟧

This code is used in section 5.3.

### 5.3.3   Reification

Reification is the transition from a valid abstract version, with operations working on an abstract state, to a valid concrete version, with operations working on a concrete state. Every concrete state is related to a unique abstract state via a *retrieve* function.

**5.3.3.1.** We will express the fact that a concrete operation $cop$ refines an abstract operation $aop$ relative to a concrete invariant $cinv$ and a retrieve function $retr$ by $cop \sqsubseteq^{op}_{cinv,retr} aop$. Formally:

⟨ Reification of operations. 5.3.3.1 ⟩ ≡
$(\cdot) \sqsubseteq^{op}_{(\cdot),(\cdot)} (\cdot) :=$
$[cstate, astate, in, out\ ?\ sort$
$\vdash [\ cop\ \ :\ OP\,(in, out, cstate);\ aop\ :\ OP\,(in, out, astate)$
$;\ cinv\ :\ [cstate \vdash prop\,];\qquad retr\ :\ [cstate \vdash astate\,]$
$\vdash [\,i\ ?\ in\ ;\ cst_i\ ?\ cstate\ ;\ ast_i\ :=\ retr\,(cst_i)$

$\vdash \dfrac{cinv\,(cst_i);\ aop(i, ast_i).\ pre}{⟨\text{Reification condition. 5.3.3.2}⟩}$

$]$
$]$
$]$

This code is used in section 5.3.3.6.

**5.3.3.2.** For every operation $aop$ of the abstract version there must be a corresponding operation $cop$ of the concrete version which satisfies the following operation reification conditions:

– The precondition must be preserved, when going from the abstract state to the concrete state (*domain*).

– The postcondition must be preserved, when going from the concrete state
to the abstract state (*result*).

⟨ Reification condition. 5.3.3.2 ⟩ ≡

◁ *domain* := *cop* ($i$, $cst_i$). *pre*

, *result* := [ $o$ ? *out* ; $cst_o$ ? *cstate* ; $ast_o$ := *retr* ($cst_o$)

$$\vdash \frac{cop\,(i, cst_i).\;post\,(o, cst_o)}{aop\,(i, ast_i).\;post\,(o, ast_o)}$$

]

▷

This code is used in section 5.3.3.1.

**5.3.3.3.** The operation reification condition is lifted to a version reification
condition, by requiring for the $n$-th operation of the abstract version to satisfy the operation reification condition wrt. the $n$-th operation of the concrete
version.

⟨ Reification of versions. 5.3.3.3 ⟩ ≡

⟦ *version_reif*(($\cdot$), ($\cdot$), ($\cdot$)) :

[*cstate*, *astate* ? *sort* ; *cinv* ? [*cstate* ⊢ *prop*]; *ainv* ? [*astate* ⊢ *prop*]

⊢[ *version* (*cinv*); *version*(*ainv*); [*cstate* ⊢ *astate*] ⊢ *prop*]

]

; *def_version_reif* :

[ *cstate*, *astate* ? *sort* ; *cinv* ? [*cstate* ⊢ *prop*]; *ainv* ? [*astate* ⊢ *prop*]

; *retr* ? [*cstate* ⊢ *astate*]

⊢◁ *empty* := *version_reif*([] ∴ *version*(*cinv*), [] ∴ *version*(*ainv*), *retr*)

, *cons* := [ *in*, *out* ? *sort*

; *cop* ? $OP$ (*in*, *out*, *cstate*); *aop* ? $OP$ (*in*, *out*, *astate*)

; *cv* ? *version* (*cinv*); *av* ? *version* (*ainv*)

$$\vdash \frac{version\_reif(cop \odot cv,\, aop \odot av,\, retr)}{◁\, base := version\_reif(cv, av, retr)}$$

◁ *base* := *version_reif*(*cv*, *av*, *retr*)

, *newop* := *cop* $\sqsubseteq^{op}_{cinv,retr}$ *aop*

▷

]

▷

]

⟧

This code is used in section 5.3.3.6.

**5.3.3.4.** Wrt. the invariants *ainv* and *cinv* of the abstract version and the
concrete version, the retrieve function has to satisfy the following correctness
conditions (*retrieve*):

- The retrieve function preserves the invariants (*preservation*).
- The retrieve function is complete, in the sense that it reaches all abstract
  states satisfying the abstract invariant from concrete states satisfying the
  concrete invariant (*completeness*).

⟨ Proof obligations for the retrieve function. 5.3.3.4 ⟩ ≡

*valid_retrieve* :=

[*cstate, astate* ? *sort*

⊢[ *cinv* : [*cstate* ⊢ *prop*]; *ainv* : [*astate* ⊢ *prop*]; *retr* : [*cstate* ⊢ *astate*]

$$\vdash \Big\langle preservation := \Big[ cst ? cstate \vdash \frac{cinv\,(cst)}{ainv\,(retr(cst))} \Big]$$

$$, completeness := \Big[ ast ? astate \vdash \frac{ainv\,(ast)}{\exists [\, cst : cstate \vdash retr(cst) = ast \wedge cinv(cst) \,]} \Big] \Big\rangle$$

⟩

]

]

This code is used in section 5.3.3.6.

**5.3.3.5.** Finally, all these conditions can be put together to define the predicate
*cv* ⊑$_{retr}$ *av*, which states that the concrete version *cv* is a reification of the
abstract version *av*, where both are related by the retrieve function *retr*.

⟨ Definition of VDM reification. 5.3.3.5 ⟩ ≡

(·) ⊑$_{(·)}$ (·) :=

[*cstate, astate* ? *sort* ; *cinv* ? [*cstate* ⊢ *prop*]; *ainv* ? [*astate* ⊢ *prop*]

⊢[ *cv* : *version* (*cinv*); *av* : *version* (*ainv*); *retr* : [*cstate* ⊢ *astate*]

⊢⟨ *version*       := ⟨ *concrete* := *cv* ✓ , *abstract* := *av* ✓ ⟩

, *retrieval*    := *valid_retrieve* (*cinv, ainv, retr*)

, *reification* := *version_reif*(*cv, av, retr*)

⟩

]

]

This code is used in section 5.3.3.6.

**5.3.3.6.** To conclude the formalization of the general theory of VDM, the given pieces can be assembled into a context formalizing VDM reification.

⟨VDM version reification. 5.3.3.6⟩ ≡
⟦⟨Reification of operations. 5.3.3.1⟩
; ⟨Reification of versions. 5.3.3.3⟩
; ⟨Proof obligations for the retrieve function. 5.3.3.4⟩
; ⟨Definition of VDM reification. 5.3.3.5⟩
⟧

This code is used in section 5.3.

## 5.4 The Human Leukocyte Antigen Case Study

### 5.4.1 Presentation

The Human-Leukocyte-Antigen typing problem stems from biology. The human organism is endowed with complex protection mechanisms. The principle of immunity is the following: as soon as an external body, e.g. an organ transplant, enters the organism, the latter will try to produce antibodies able to recognize this stranger, with the intention of destroying it.

The recognition of a cell by an antibody occurs when the given cell contains an antigen corresponding to the antibody. This reaction results in destruction of the cell. The ability to foresee such reactions is very important in the medical science, for example for organ transplants. In fact, it has been observed that each human cell has the same combination of antigens (*Human-Leukocyte-Antigen* complex). "HLA typing" means determining this set of antigens. HLA typing is realized by bringing cells (containing antigens) and sera (containing antibodies) together and observing their reactions.

The HLA complex and HLA typing is described, for instance, in [37], where we have taken the following quotation (p. 382ff):

> The human leukocyte antigen (HLA) complex, the human version
> of the major histocompatibility complex, is a cluster of genes on chro-
> mosome 6 which codes for a unique set of proteins on the surfaces of
> most cells. These proteins have the ability to bind antigenic fragments
> of other proteins made in or entering the same cell and present them to
> the T-cell receptor of T lymphocytes. [...] The presentation initiates the
> specific immune-response in the human body.
>
> Because different individuals have different variants of HLA molecules
> and because each variant can only bind some protein fragments, individ-
> uals differ in their ability to respond to different proteins. If a particular
> variant of the HLA molecules fails to bind a particular fragment, this
> individual's T lymphocytes ignore the corresponding antigen and fail to
> initiate a specific immune response to it, while responding normally to

*other antigens. [...] The HLA complex thus functions as a set of im-
mune response genes, deciding on the antigenic determinants to which a
person can respond. [...]*

*In a randomly assembled sample of individuals, the chances are that
no two individuals will have completely identical HLA molecules. This
fact has grave consequences for attempts to transplant organs between
individuals. The recipient's T lymphocytes recognize the different HLA
molecules on the donor's organ as foreign and become strongly stimulated
by them. They then initiate an immune response that can lead to the
rejection (i.e., destruction) of the transplant. In an effort to minimize
the antigenic stimulus to the recipient's lymphocytes, the donor and the
recipient can be HLA typed and matched for at least some of the antigens.*

*This matching can be relatively good between members of the same
family, but complete matching between unrelated individuals is usually
not possible. The HLA typing is carried out in specialized laboratories
with batteries of antibodies with considerable specificity for the individual
HLA antigens.*

The HLA typing process could be compared to a "Master Mind" game where
the solution would correspond to the unknown serum and the preliminary trials
(with their associated results), to the data of the HLA typing process, that is the
experiments of reactions between the serum and a succession of cells. Conversely,
it could be the cell that is unknown.

The example presented describes the typing of an unknown serum by reaction
with different cells. The dual problem will not be treated here.

*Warning:* Do not confuse HLA "typing" with Deva "typing"!

### 5.4.2   Development in VDM

This section proposes a VDM formalization of the HLA typing problem. The
VDM notations are taken from [60]. The one-to-one correspondence between
antigens and antibodies suggests that there will be no need to have separate
kinds of sets for these two concepts. A single kind of set (antigen or $\mathcal{A}$) is used,
the context will determine which is which.

### VDM specification

Let $\mathcal{A}$ be the set of antigens and $\mathcal{A}_{cred}$ the set of credible antigens, defined by
using a credibility predicate *cred* (credible means a high probability of reaction).
Two results are possible for an experiment putting together a cell and the
unknown serum: positive or negative. An experiment is made up of a cell and
its corresponding result. The VDM expression of these concepts follows:

$$\mathcal{A} = \{x_1, \ldots, x_n\}$$

$cred : \mathcal{A} \rightarrow \mathbb{B}$

$cred(a) \quad \triangleq \quad \ldots$

$\mathcal{A}_{cred} = \{a \in \mathcal{A} \mid cred(a)\}$

$Result = \{+, -\}$

$Exper \;::\; Cell \;:\; \mathcal{A}\text{-set}$
$\qquad\qquad Res \;:\; Result$

The global information (VDM state) consists of a set of experiments (given as data) and of three sets of cells (to be constructed by the different operations). These are the sets of cells reacting positively ($Pos$) or negatively ($Neg$) and the set of cells generating an inconsistency ($Fail$). After successive transformations, $Pos$ will contain the final result: a set of sets of antigens where each set of antigens is a possible resulting serum. No special constraints are imposed on this state:

$State_{HLA} \;::\; \qquad Resexp \;:\; Exper\text{-set}$
$\qquad\qquad Pos, Neg, Fail \;:\; \mathcal{A}\text{-set-set}$

where

$inv\_State_{HLA}(s) \quad \triangleq \quad$ true

A set of operations on the state is defined determining the serum. These operations may be considered as the result of an operation decomposition step; this explains why they make sense only if applied successively. Remember, that an operation is given by a pre- and a postcondition, if the precondition is true it can be omitted. Moreover, variables to which the operation has external access may be declared read-and-write (extwr) or read-only (extrd).

$INIT$ initializes the state.

$INIT \;(Resexp_0 : Exper\text{-set})$
ext  wr $Resexp, Pos, Neg, Fail$
post  $Resexp = Resexp_0 \wedge Pos, Neg, Fail = \{\,\}$

$DETPOSNEG$ generates two sets of cells from $Resexp$, by projection, according to the value of the reaction (+ or −) of this cell with the unknown serum.

$DETPOSNEG$
ext  rd $Resexp, Fail$
$\qquad$ wr $Pos, Neg$
post  $Pos = \{p \in \mathcal{A}\text{-set} \mid mk\_Exper(p, +) \in Resexp\}$
$\qquad \wedge$
$\qquad\quad Neg = \{p \in \mathcal{A}\text{-set} \mid mk\_Exper(p, -) \in Resexp\}$

*DETFAIL* detects inconsistencies; when all the antigens of a positive cell are also in one or several negative cells, this positive cell generates a contradiction and is called "head of inconsistency".

> *DETFAIL*
> ext  rd *Resexp, Pos, Neg*
>      wr *Fail*
> post  $Fail = \{p \in Pos \mid \forall x \in p \cdot \exists m \in Neg \cdot x \in m\}$

Inconsistencies are eliminated by the next three operations.

*ELIMPOS* eliminates from *Pos* heads of inconsistencies that do not contain any credible antigen (overlined variables symbolize their value before the operation).

> *ELIMPOS*
> ext  rd *Resexp, Neg, Fail*
>      wr *Pos*
> post  $Pos = \{p \in \overleftarrow{Pos} \mid (p \cap \mathcal{A}_{cred} = \{\}) \Rightarrow p \notin Fail\}$

*ELIMNEG* eliminates from *Neg* negative cells including at least a credible antigen also belonging to a head of inconsistency.

> *ELIMNEG*
> ext  rd *Resexp, Pos, Fail*
>      wr *Neg*
> post  $Neg = \{p \in \overleftarrow{Neg} \mid (p \cap \mathcal{A}_{cred} = \{\}) \Rightarrow p \notin Fail\}$

*ELIMPOS2* eliminates from *Pos* remaining positive cells which are heads of inconsistencies not yet eliminated.

> *ELIMPOS2*
> ext  rd *Resexp, Neg*
>      wr *Pos, Fail*
> post  $Fail = \{p \in \overleftarrow{Pos} \mid \forall x \in p \cdot \exists m \in Neg \cdot x \in m\}$
>       $\wedge\, Pos = \overleftarrow{Pos} \setminus Fail$

*DETSER* imposes that the final result (recorded in *Pos*) is the minimal set of possible sera.

> *DETSER*
> ext  rd *Resexp, Neg, Fail*
>      wr *Pos*
> post  $Pos = \{q \in \mathcal{A}\text{-set} \mid \exists p \in \overleftarrow{Pos} \cdot q = (p \setminus \bigcup \overleftarrow{Neg})\}$

**VDM reification**

The VDM reification aims to transform the initial abstract state $(State_{HLA})$ into an efficiently implementable data structure. The target language chosen in [25] was LISP; this suggests the choice of lists (sequences) as implementation of sets. Other solutions have also been explored in [21]: bignums and balanced trees.

So, the concrete state is built from lists and the invariant requires the absence of duplicates in lists at all levels (list of cells and cells).

$$cExper \; :: \; cCell \; : \; \mathcal{A}^*$$
$$cRes \; : \; Result$$

$$cState_{HLA} \; :: \qquad cResexp \; : \; cExper^*$$
$$cPos, cNeg, cFail \; : \; \mathcal{A}^{**}$$

where

$$inv_c State_{HLA}(cs) \; \triangle$$
$$noDupl(cPos(cs)) \land noDupl(cNeg(cn)) \land noDupl(cFail(cf))$$

The "no-duplicate" condition can be expressed requiring, for example, that the cardinality of the set of elements contained in these lists is equal to the length of these lists. This condition must hold at the different levels (antigens and cells).

$$noDupl : \mathcal{A}^{**} \rightarrow \mathbb{B}$$

$$noDupl(ll) \; \triangle \quad \mathsf{card\,elems}\, ll = \mathsf{len}\, ll$$
$$\land \, \forall l \in \mathcal{A}^* \cdot l \in \mathsf{elems}\, ll \; \Rightarrow \; \mathsf{card\,elems}\, l = \mathsf{len}\, l$$

The "no-duplicate" has already been defined in Chap. 4, and the reader may find it interesting to compare these definitions.

$$no\_dupl := [\, s \; ? \; sort \,; \; xs \; : \; seq \, (seq(s))$$
$$\vdash card \, (elems(xs)) = len(xs)$$
$$\land \, \forall [\, x : seq \, (s) \vdash x \in elems(xs) \Rightarrow card(elems(x)) = len(x)\,]$$
$$\,]$$

The retrieve function returns the abstract state corresponding to a given concrete state. Here, it transforms each sequence into the set of its elements. Once more the difficulty lies in the nesting of the sequences.

$$Retr_1 : cExper^* \rightarrow cExper\text{-set}$$

$$Retr_1(l) \; \triangle \quad \text{if} \; l = []$$
$$\text{then} \; \{\}$$
$$\text{else} \; (\text{elems}\, cCell(\mathsf{hd}\, l) \mapsto cRes(\mathsf{hd}\, l) \odot Retr_1(\mathsf{tl}\, l))$$

$Retr_2 : \mathcal{A}^{**} \to \mathcal{A}\text{-set-set}$

$Retr_2(ll) \quad \triangleq \quad$ if $ll = []$
   then $\{\}$
   else elems hd $l \odot Retr_2(\text{tl } ll)$

$Retr_{HLA} : cState_{HLA} \to State_{HLA}$

$Retr_{HLA}(cs) \quad \triangleq$
   $mk-State_{HLA}(Retr_1(cResexp(cs)), Retr_2(cPos(cs)),$
   $Retr_2(cNeg(cs)), Retr_2(cFail(cs)))$

The operations are adapted to the new data structure and their postcondition is made more functional. This is illustrated on the *DETPOSNEG* operation This operation will be used as "guinea-pig" in the following formalization of the development.

$cellfilter : Result \times cExper^* \to \mathcal{A}^{**}$

$cellfilter(r, l) \quad \triangleq \quad$ if $cRes(\text{hd } l) = r$
   then $cons(\text{hd } l, cellfilter(r, \text{tl } l))$
   else $cellfilter(r, \text{tl } l)$

*cDETPOSNEG*

ext  rd $cResexp, cFail$
     wr $cPos, cNeg$

post $cPos = cellfilter(+, \overleftarrow{cResexp})$
   $\land cNeg = cellfilter(-, \overleftarrow{cResexp})$

## 5.5 Formalization of the HLA Development in Deva

The HLA typing problem was presented and described in VDM in Sect. 5.4. This section will illustrate the Deva formalization of VDM reification presented in Sect. 5.3 by formalizing the reification step of the HLA development. The HLA development presentation is structured as follows:

- First, some global primitives of the HLA case study are introduced.
- Then, the *state descriptions*, i.e., state, invariant, and operations, of the abstract version $Ver_{HLA}$ and the concrete version $cVer_{HLA}$ of the HLA specification, and the retrieve function $Retr_{HLA}$ are defined.
- Finally, the *correctness* of the reification step will be verified by formally proving the reification relation between the abstract state and the concrete state.

The general structure of the Deva formalization reflects this organization:

⟨Human leukocyte antigen typing problem. 5.5⟩ ≡

**context** *HLACaseStudy* :=

⟦⟨HLA global parameters. 5.5.1⟩
;⟨HLA abstract specification. 5.5.2.9⟩
;⟨HLA concrete specification. 5.5.3.4⟩
;⟨HLA retrieve function. 5.5.4.6⟩
;⟨HLA verification. 5.5.6⟩
⟧

This code is used in section 5.1.

### 5.5.1  HLA Primitives

The global primitives of the development are formalized as global parameters:
The sort $\mathcal{A}$ of antigens with a credibility predicate, and the sort *Result* of result
indications, which are either $+$ or $-$.

⟨HLA global parameters. 5.5.1⟩ ≡

⟦$\mathcal{A}$                      : *sort*
; *cred*              : $[\mathcal{A} \vdash prop]$
; *Result*            : *sort*
; $+, -$              : *Result*
; *no_other_results* : $[r\,?\,Result \vdash r = +\vee r = -]$
⟧

This code is used in section 5.5.

### 5.5.2  HLA Abstract Specification

The Deva specification of the State of the HLA problem is very close to the
VDM description (Sect. 5.4). An experiment is defined as a couple, i.e., a pair
of elements of type sort. *Cell* and *Res* are selectors on experiments. One may
observe that Deva allows an explicit link between the invariant and the state it
is attached to. This link is contained in the type of the invariant. In VDM, only
the identifier of the invariant contained a reference to the corresponding state.
This shows how Deva helps to express the structure of VDM objects.

⟨HLA abstract specification of the state and the invariant. 5.5.2⟩ ≡

⟦ *Exper*    := $T(set(\mathcal{A}) \otimes Result \otimes mt)$
; *Cell*     := $sel\,.car\,\therefore [Exper \vdash set(\mathcal{A})]$
; *Res*      := $[\,e : Exper \vdash sel.\,car\,(sel.\,cdr\,(e))\,]$

; $State_{HLA} := T(set(Exper) \otimes set(set(\mathcal{A})) \otimes set(set(\mathcal{A})) \otimes set(set(\mathcal{A})) \otimes mt)$
; $Resexp \quad := [\, st : State_{HLA} \vdash sel.\, car\,(st)\,]$
; $Pos \qquad := [\, st : State_{HLA} \vdash sel.\, car\,(sel.\, cdr\,(st))\,]$
; $Neg \qquad := [\, st : State_{HLA} \vdash sel.\, car\,(sel.\, cdr\,(sel.\, cdr\,(st)))\,]$
; $Fail \qquad := [\, st : State_{HLA} \vdash sel.\, car\,(sel.\, cdr\,(sel.\, cdr\,(sel.\, cdr\,(st))))\,]$
; $Inv_{HLA} \quad := [State_{HLA} \vdash true\,]$
$\rrbracket$

This code is used in section 5.5.2.9.

**5.5.2.1.**  Some useful selection laws about the abstract state are derived using a tactic ($tuple\_unf\_tactic$) to unfold selections from tuples.

$\langle$ Derivation of selection laws. 5.5.2.1 $\rangle \equiv$
$\llbracket$ $sel_{Res} \quad := [\, c\; ?\; set\,(\mathcal{A}); \; r\; ?\; Result$
$\qquad\qquad \vdash refl$
$\qquad\qquad\qquad \therefore Res(c \mapsto r) = sel.\, car\,(sel.\, cdr\,(c \mapsto r))$
$\qquad\qquad \setminus \textbf{loop}\; tuple\_unf\_tactic$
$\qquad\qquad\qquad \therefore Res(c \mapsto r) = r$
$\qquad\qquad ]$
; $sel_{HLA} := [\, R\; ?\quad set\,(Exper); P, N, F\; ?\; set\,(set(\mathcal{A}))$
$\qquad\qquad ; \; st := mk\,(R, mk(P, mk(N, mk(F, mkmt))))$
$\qquad\qquad \vdash\!\langle\!\langle Resexp := \langle Resexp\; \text{selection.}\; 5.5.2.2\,\rangle$
$\qquad\qquad\quad , Pos \qquad := \langle Pos\; \text{selection.}\; 5.5.2.3\,\rangle$
$\qquad\qquad\quad , Neg \qquad := \langle Neg\; \text{selection.}\; 5.5.2.4\,\rangle$
$\qquad\qquad\quad , Fail \qquad := \langle Fail\; \text{selection.}\; 5.5.2.5\,\rangle$
$\qquad\qquad\quad \rangle\!\rangle$
$\qquad\qquad ]$
$\rrbracket$

This code is used in section 5.5.2.9.

**5.5.2.2.**    $\langle$ Resexp selection. 5.5.2.2 $\rangle \equiv$
$refl$
$\quad \therefore Resexp(st) = sel.\, car\,(st)$
$\setminus \textbf{loop}\; tuple\_unf\_tactic$
$\quad \therefore Resexp(st) = R$

This code is used in section 5.5.2.1.

**5.5.2.3.**     ⟨ *Pos* selection. 5.5.2.3 ⟩ ≡

*refl*

   ∴ *Pos*(*st*) = *sel. car* (*sel. cdr* (*st*))

\ **loop** *tuple_unf_tactic*

   ∴ *Pos*(*st*) = *P*

This code is used in section 5.5.2.1.

**5.5.2.4.**     ⟨ *Neg* selection. 5.5.2.4 ⟩ ≡

*refl*

   ∴ *Neg*(*st*) = *sel. car* (*sel. cdr* (*sel. cdr* (*st*)))

\ **loop** *tuple_unf_tactic*

   ∴ *Neg*(*st*) = *N*

This code is used in section 5.5.2.1.

**5.5.2.5.**     ⟨ *Fail* selection. 5.5.2.5 ⟩ ≡

*refl*

   ∴ *Fail*(*st*) = *sel. car* (*sel. cdr* (*sel. cdr* (*sel. cdr* (*st*))))

\ **loop** *tuple_unf_tactic*

   ∴ *Fail*(*st*) = *F*

This code is used in section 5.5.2.1.

**5.5.2.6.**     According to the syntax given in Sect. 5.3.1, the operation determining positive and negative cells (our guinea pig operation) is now specified. Positive and negative sets are built by filtering the initial set of cells according to their result in reactions. See Chap. 4.3.2 for the specification of the filter operation on sets. *DetPosNeg* does not have any input or output parameters, it only transforms the state. Hence we use the dummy type *void* as parameter type. The formalization is shown in Fig. 13 where it is shown together with the VDM definition given earlier. The difference in size results mainly from the fact that the Deva definition does not use a number of notational shortcuts of VDM. These shortcuts are very important though and it could be an interesting and useful experiment to try to program a translator that is able to generate the Deva version given the VDM version, and vice versa.

**5.5.2.7.**     We are not interested at this point in the specifications of the other operations. Therefore, instead of *defining* them, as it was done for *DetPosNeg*, we just *declare* their types.

⟨ Declaration of the other abstract operations. 5.5.2.7 ⟩ ≡

⟦ *Init*     : *OP* (*set*(*Exper*), *void*, *State*$_{HLA}$)

$\langle$ Specification of *DetPosNeg.* 5.5.2.6 $\rangle \equiv$

*DetPosNeg* :=

$[\, IP \,:\, void \,;\, I \,:\, State_{HLA}$

$\vdash\!\{\, pre \;:=\; true$

$,\ post \;:= [\, OP \,:\, void \,;\, O \,:\, State_{HLA}$

$\qquad\qquad \vdash Pos\,(O) = Cell * [\, e : Exper \vdash Res(e) = +\,] \,\triangleright\, Resexp(I)$

$\qquad\qquad \wedge\, Neg(O) = Cell * [\, e : Exper \vdash Res(e) = -\,] \,\triangleright\, Resexp(I)$

$\qquad\qquad \wedge\, Fail(O) = Fail(I)$

$\qquad\qquad \wedge\, Resexp(O) = Resexp(I)$

$\qquad\quad ]$

$\quad \flat$

$]$

$\qquad \therefore OP(void, void, State_{HLA})$

This code is used in section 5.5.2.9.

---

*DETPOSNEG*

**ext** **rd** *Resexp, Fail*

$\qquad$ **wr** *Pos, Neg*

**post** $Pos = \{p \in \mathcal{A}\text{-set} \mid mk\_Exper(p, +) \in Resexp\}$

$\qquad\ \wedge$

$\qquad Neg = \{p \in \mathcal{A}\text{-set} \mid mk\_Exper(p, -) \in Resexp\}$

**Fig. 13.** The operation *DETPOSNEG* in Deva and in VDM.

; *DetFail* $\quad : OP\,(void, void, State_{HLA})$

; *ElimPos* $\quad : OP\,(void, void, State_{HLA})$

; *ElimNeg* $\quad : OP\,(void, void, State_{HLA})$

; *ElimPos2* $: OP\,(void, void, State_{HLA})$

; *DetSer* $\quad : OP\,(void, void, State_{HLA})$

$]$

This code is used in section 5.5.2.9.

Methodically, these declarations raise an interesting point, because they could be considered as yet to-be-satisfied "specification obligations" of an incomplete specification. Such a specification can already be sufficient to begin with first parts of the formal development. In fact, the reader will see precisely how this is possible. Of course, in a more complete presentation (which is outside the scope of this book), more (and eventually all) operations could be specified, and, given this new information, more useful properties could be deduced.

The decision how much to specify and when to start with formal proof depends on the objectives of the development activity. This illustrates again that the purpose of Deva is not to impose a methodological strait-jacket but to be a notation for the "logical bookkeeping" of the developers activities.

**5.5.2.8.** The operations are combined into the abstract version.

$\langle$ Construction of the abstract version. 5.5.2.8 $\rangle \equiv$

$Ver_{HLA} :=$

$DetSer \circledcirc ElimPos2 \circledcirc ElimNeg \circledcirc ElimPos \circledcirc DetFail \circledcirc DetPosNeg \circledcirc Init \circledcirc[]$

$\quad \therefore version(Inv_{HLA})$

This code is used in section 5.5.2.9.

**5.5.2.9.** By combining all contexts of this section, one obtains the abstract specification of the HLA typing problem.

$\langle$ HLA abstract specification. 5.5.2.9 $\rangle \equiv$

$[\![\langle$ HLA abstract specification of the state and the invariant. 5.5.2 $\rangle$

$; \langle$ Derivation of selection laws. 5.5.2.1 $\rangle$

$; \langle$ Specification of $DetPosNeg$. 5.5.2.6 $\rangle$

$; \langle$ Declaration of the other abstract operations. 5.5.2.7 $\rangle$

$; \langle$ Construction of the abstract version. 5.5.2.8 $\rangle$

$]\!]$

This code is used in section 5.5.

### 5.5.3  HLA Concrete Specification

The concrete version of the state is very close to the abstract one, each set is replaced by a sequence. The concrete invariant requires that the sequences contain no duplicates.

$\langle$ HLA concrete specification of the state and the invariant. 5.5.3 $\rangle \equiv$

$[\![\ cExper \quad\ := T(seq(\mathcal{A}) \otimes Result \otimes mt)$

$; \ cCell \quad\quad := [\, e : cExper \vdash sel. \, car\,(e)\,]$

$; \ cRes \quad\quad := [\, e : cExper \vdash sel. \, car\,(sel. \, cdr\,(e))\,]$

$; \ cState_{HLA} :=$

$\quad T(seq(cExper) \otimes seq(seq(\mathcal{A})) \otimes seq(seq(\mathcal{A})) \otimes seq(seq(\mathcal{A})) \otimes mt)$

$; \ cResexp \quad := [\, st : cState_{HLA} \vdash sel. \, car\,(st)\,]$

; $cPos$           $:= [\, st : cState_{HLA} \vdash sel.\, car\,(sel.\, cdr\,(st))\,]$
; $cNeg$           $:= [\, st : cState_{HLA} \vdash sel.\, car\,(sel.\, cdr\,(sel.\, cdr\,(st)))\,]$
; $cFail$          $:= [\, st : cState_{HLA} \vdash sel.\, car\,(sel.\, cdr\,(sel.\, cdr\,(sel.\, cdr\,(st))))\,]$
; $cInv_{HLA}$     $:= [\, cs\ :\ cState_{HLA}$
                    $\qquad \vdash no\_dupl\,(cPos(cs)) \wedge no\_dupl(cNeg(cs)) \wedge no\_dupl(cFail(cs))$
                    $\qquad ]$

$]\!]$

This code is used in section 5.5.3.4.

**5.5.3.1.** The operation $cDetPosNeg$ is implemented using the auxiliary function $cellfilter$. This function filters out all those cells from a sequence of experiments, which react either positively or negatively.

$\langle$ Specification of $cDetPosNeg$. 5.5.3.1 $\rangle \equiv$
$[\![\ cellfilter\quad := [\, r\ :\ Result\,;\ ll\ :\ seq\,(cExper)$
                    $\qquad \vdash cCell *_{seq}([\, e : cExper \vdash cRes(e) = r\,] \triangleright_{seq} ll)$
                    $\qquad ]$
; $cDetPosNeg := [\, IP\ :\ void\,;\ I\ :\ cState_{HLA}$
                $\qquad \vdash\!\langle\ pre\ :=\ true$
                $\qquad ,\ post\ := [\, OP\ :\ void\,;\ O\ :\ cState_{HLA}$
                $\qquad\qquad \vdash cPos\,(O) = cellfilter(+, cResexp(I))$
                $\qquad\qquad \wedge\ cNeg(O) = cellfilter(-, cResexp(I))$
                $\qquad\qquad \wedge\ cFail(O) = cFail(I)$
                $\qquad\qquad \wedge\ cResexp(O) = cResexp(I)$
                $\qquad\qquad ]$
                $\qquad \rangle$
                $\qquad ]$
                $\qquad\qquad \therefore\ OP(void, void, cState_{HLA})$

$]\!]$

This code is used in section 5.5.3.4.

**5.5.3.2.** Similar to the abstract version, the other operations are not defined but declared. Note that a developer could very well define a concrete operation, say $cDetFail$, at this point, and later recover from that the abstract definition $DetFail$.

$\langle$ Declaration of the other concrete operations. 5.5.3.2 $\rangle \equiv$

$$\begin{aligned}
[\![ \ cInit \quad &: \ OP\,(set(Exper), void, cState_{HLA}) \\
; \ cDetFail \quad &: \ OP\,(void, void, cState_{HLA}) \\
; \ cElimPos \quad &: \ OP\,(void, void, cState_{HLA}) \\
; \ cElimNeg \quad &: \ OP\,(void, void, cState_{HLA}) \\
; \ cElimPos2 \quad &: \ OP\,(void, void, cState_{HLA}) \\
; \ cDetSer \quad &: \ OP\,(void, void, cState_{HLA}) \\
]\!]
\end{aligned}$$

This code is used in section 5.5.3.4.

**5.5.3.3.** The operations are combined into the concrete version.

$\langle$ Construction of the concrete version. 5.5.3.3 $\rangle \equiv$

$$\begin{aligned}
cVer_{HLA} := \ &cDetSer \odot cElimPos2 \odot cElimNeg \odot cElimPos \odot \\
&\quad cDetFail \odot cDetPosNeg \odot cInit \odot [\!] \\
&\quad \therefore version(cInv_{HLA})
\end{aligned}$$

This code is used in section 5.5.3.4.

**5.5.3.4.** By combining all contexts of this section, one obtains the concrete specification of the HLA typing problem.

$\langle$ HLA concrete specification. 5.5.3.4 $\rangle \equiv$
$[\![ \langle$ HLA concrete specification of the state and the invariant. 5.5.3 $\rangle$
$; \langle$ Specification of $cDetPosNeg$. 5.5.3.1 $\rangle$
$; \langle$ Declaration of the other concrete operations. 5.5.3.2 $\rangle$
$; \langle$ Construction of the concrete version. 5.5.3.3 $\rangle$
$]\!]$

This code is used in section 5.5.

### 5.5.4 Specification of the Retrieve Function

The retrieve function is built up in two stages. The first part $(Retr_1)$ concerns the set of experiments $cResexp$, the second part $(Retr_2)$ treats the sets of antigen sets $cPos$, $cNeg$, and $cFail$. Both are defined in a recursive manner, transforming each sequence into the set of its elements, using the function $elems$. In comparison with the VDM definitions given above, the case distinctions have been split into individual cases within a named product. This allows to refer to the particular cases within deductions by name.

⟨ Two auxiliary specifications. 5.5.4 ⟩ ≡
⟦ $Retr_1$        : [ $seq\,(cExper) \vdash set(Exper)$ ]
; $Retr_1\_def$ : [ $e$ ? $cExper$ ; $x$ ? $seq\,(cExper)$
                    $\vdash\!\langle\ base := Retr_1\,(\langle\rangle) = \{\}$
                    , $rec$   := $Retr_1\,(e :: x)$
                                = $(elems(cCell(e)) \mapsto cRes(e)) \odot Retr_1(x)$
                    ⟩
                ]
; $Retr_2$        : [ $seq\,(seq(\mathcal{A})) \vdash set(set(\mathcal{A}))$ ]
; $Retr_2\_def$ : [ $x$ ? $seq\,(\mathcal{A})$; $xs$ ? $seq\,(seq(\mathcal{A}))$
                    $\vdash\!\langle\ base := Retr_2\,(\langle\rangle) = \{\}$
                    , $rec$   := $Retr_2\,(x :: xs) = elems(x) \odot Retr_2(xs)$
                    ⟩
                ]
⟧

This code is used in section 5.5.4.6.


**5.5.4.1.**   The retrieve function is constructed as its VDM counterpart (p. 141).

⟨ Definition of the retrieve function. 5.5.4.1 ⟩ ≡
$Retr_{HLA}$ := [ $cst$ : $cState_{HLA}$
                    $\vdash mk\,(Retr_1(cResexp(cst)), mk(Retr_2(cPos(cst)),$
                       $mk\,(Retr_2(cNeg(cst)), mk(Retr_2(cFail(cst)), mkmt))))$
                ]

This code is used in section 5.5.4.6.


**5.5.4.2.**   For later use, the defining laws for the retrieve function are summarized into a small tactic, and this tactic is then added to the VDM evaluation tactic defined in Sect. 4.3.6 to yield a tactic for evaluation in the context of the HLA development.

**5.5.4.3.**   ⟨ Tactic for evaluating the retrieve function. 5.5.4.3 ⟩ ≡
$RetrEvaluation$ :=
⟨ $base$ := **alt** $[rsubst(Retr_1\_def.base), rsubst(Retr_2\_def.base)]$
, $rec$   := **alt** $[rsubst(Retr_1\_def.rec), rsubst(Retr_2\_def.rec)]$
⟩

This code is used in section 5.5.4.6.

**5.5.4.4.** ⟨ General evaluation tactic for the HLA case study. 5.5.4.4 ⟩ ≡
*HLAEvaluation* :=
⦇ *empty* := **alt** [ *VDMEvaluation. base* , *RetrEvaluation. base* ]
, *cons* := **alt** [ *VDMEvaluation. rec* , *RetrEvaluation. rec* ]
⦈

This code is used in section 5.5.4.6.

**5.5.4.5.** Finally, two useful properties about the retrieve functions are deduced. The presentation of the proofs is deferred to later sections.

⟨ Two retrieve lemmas. 5.5.4.5 ⟩ ≡
⟦ *retr_lemma₁* :=
  ⟨ Proof of the first retrieve lemma. 5.5.8 ⟩

$$\therefore [\, x\,?\, set\,(\mathcal{A}); l\,?\ seq\,(cExper); r\,?\ Result \vdash \frac{x \in Retr_2(cellfilter(r, l))}{(x \mapsto r) \in Retr_1(l)}\,]$$

; *retr_lemma₂* :=
  ⟨ Proof of the second retrieve lemma. 5.5.5 ⟩

$$\therefore [\, l\,?\ seq\,(cExper); r\ ?\ Result$$
$$\vdash Retr_2\,(cellfilter(r, l)) = Cell * [\, e : Exper \vdash Res(e) = r\,] \triangleright Retr_1(l)$$
$$\,]$$

⟧

This code is used in section 5.5.4.6.

**5.5.4.6.** The material of this section can be combined to yield the specification of the retrieve function.

⟨ HLA retrieve function. 5.5.4.6 ⟩ ≡
⟦ ⟨ Two auxiliary specifications. 5.5.4 ⟩
; ⟨ Definition of the retrieve function. 5.5.4.1 ⟩
; ⟨ Tactic for evaluating the retrieve function. 5.5.4.3 ⟩
; ⟨ General evaluation tactic for the HLA case study. 5.5.4.4 ⟩
; ⟨ Two retrieve lemmas. 5.5.4.5 ⟩
⟧

This code is used in section 5.5.

### 5.5.5 Proof of a Property of the Retrieve Function

The second of the two properties of the retrieve function is now formally derived. The proof introduces two global variables, defines two useful sets, makes a little

side-calculation (*aside*), and then enters its main body in which the second
retrieve lemma is derived.

⟨Proof of the second retrieve lemma. 5.5.5⟩ ≡
[ r ? Result ; l ? seq (cExper)
⊢[ Filtered_Expers := [ e : Exper ⊢ Res(e) = r] ▷ Retr₁(l)
 ; Filtered_Cells   := Cell * Filtered_Expers
 ; aside            := [ e₁ : Exper ⊢ [ hyp : e₁ ∈ Filtered_Expers ]
                                    ⊢ hyp
                                       \ props_in. filter . down
                                       \ conj. elimr
                                           ∴ Res(e₁) = r
                                    ]
⊢⟨Body of the proof of the second retrieve lemma. 5.5.5.1⟩
 ]
]

This code is used in section 5.5.4.5.

**5.5.5.1.**    The idea is to prove equality of two sets by proving that all their
elements are equal ( *observ .up*), and we prove this by a sequence of equivalence
transformations, starting off with a tautology.

⟨Body of the proof of the second retrieve lemma. 5.5.5.1⟩ ≡
[ x : set (𝒜)
⊢ true_proof
      ∴ true
  \ equiv_props. refl . up
      ∴ x ∈ Filtered_Cells ⇔ x ∈ Filtered_Cells
  \ psubst(tuple_prop(aside))
      ∴ x ∈ Filtered_Cells ⇔ (x ↦ r) ∈ [ e : Exper ⊢ Res(e) = r] ▷ Retr₁(l)
  \ psubst(props_in.filter)
      ∴ x ∈ Filtered_Cells ⇔ (x ↦ r) ∈ Retr₁(l) ∧ Res(x ↦ r) = r
  \ equiv_props. conj_right (sel_Res). up
      ∴ x ∈ Filtered_Cells ⇔ (x ↦ r) ∈ Retr₁(l)
  \ prsubst(retr_lemma₁)
      ∴ x ∈ Filtered_Cells ⇔ x ∈ Retr₂(cellfilter(r, l))
  \ equiv_props. sym . down
      ∴ x ∈ Retr₂(cellfilter(r, l)) ⇔ x ∈ Filtered_Cells
]

$\backslash\ univ.\ intro\ (P := [\,x : set\,(\mathcal{A}) \vdash x \in Retr_2(cellfilter(r, l)) \Leftrightarrow x \in Filtered\_Cells\,])$

$\backslash\ observ.\ up$

$\quad \therefore Retr_2(cellfilter(r, l)) = Filtered\_Cells$

This code is used in section 5.5.5.

It is both interesting and instructive to compare this formal proof style with the informal but rigorous style advocated for VDM [60], see also [47], [69]. The corresponding VDM proof of the second retrieve lemma is shown in Fig. 14. The reasoning in the rigorous proof is analogous to the reasoning in the Deva proof. Note, the justification of the reasoning steps is recommended but not mandatory in the rigorous proof style considered here. This example makes a compromise by indicating the laws justifying each step, but not the substitution principles, i.e., *psubst*, *prsubst*, used in some steps. The formal proof in Deva, in contrast, requires a justification of each line.

---

|  | from $r?\,Result;\ l?\,seq(cExper)$ | |
|---|---|---|
| 1 | from $e_1 \in Exper;\ e_1 \in [e : Exper \vdash Res(e) = r] \rhd Retr_1(l)$ | |
| 1.1 | $\cdots \wedge Res(e_1) = r$ | $props\_in.filter.down(h1)$ |
|  | infer $Res(e_1) = r$ | $conj.elimr(1.1)$ |
| 2 | from $x \in set(\mathcal{A})$ | |
| 2.1 | $x \in Filtered\_Cells \Leftrightarrow x \in Filtered\_Cells$ | |
|  |  | $equiv\_props.refl.up(h2)$ |
| 2.2 | $x \in Filtered\_Cells$ | |
|  | $\Leftrightarrow (x \mapsto r) \in [e : Exper \vdash Res(e) = r] \rhd Retr_1(l)$ | |
|  |  | $tuple\_prop(1, 2.1)$ |
| 2.3 | $x \in Filtered\_Cells$ | |
|  | $\Leftrightarrow (x \mapsto r) \in Retr_1(l) \wedge Res(x \mapsto r) = r$ | |
|  |  | $props\_in.filter(2.2)$ |
| 2.4 | $x \in Filtered\_Cells \Leftrightarrow (x \mapsto r) \in Retr_1(l)$ | |
|  |  | $equiv\_props.conj\_right(sel_{Res}).up(2.3)$ |
| 2.5 | $x \in Filtered\_Cells \Leftrightarrow x \in Retr_2(cellfilter(r, l))$ | |
|  |  | $retr\_lemma1(2.4)$ |
| 2.6 | $x \in Retr_2(cellfilter(r, l)) \Leftrightarrow x \in Filtered\_Cells$ | |
|  |  | $equiv\_props.sym.down(2.5)$ |
|  | infer $x \in Retr_2(cellfilter(r, l)) \Leftrightarrow x \in Filtered\_Cells$ | |
|  |  | $equiv\_props.sym.down(2.5)$ |
| 3 | $\forall x : set(\mathcal{A}) \cdot x \in Retr_2(cellfilter(r, l)) \Leftrightarrow x \in Filtered\_Cells$ | |
|  |  | $univ.intro(2)$ |
|  | infer $Retr_2(cellfilter(r, l)) = Filtered\_Cells$ | $observ.up(3)$ |

**Fig. 14.** A VDM proof of the second retrieve lemma.

Note, the rigorous version is split into three main parts listed and numbered sequentially. This somewhat hides the overall structure of the deduction, which is reflected clearer through the block-structuration used in the formal version. Also, somewhat surprisingly, the formal proof does not need integer-labels and integer-pointers.

### 5.5.6   HLA Development Construction

Developing the HLA problem consists of two parts. The first part described in the previous sections was to specify the abstract version $Ver\_HLA$, the concrete version $cVer\_HLA$, and the retrieve function $Retr\_HLA$. This section describes the second part which is to construct a proof verifying the reification relation between these two versions. Since this verification can be seen very naturally as based on a set of partially discharged proof obligations, we have chosen the following overall scheme of the proof:

$\langle$ HLA verification. 5.5.6 $\rangle \equiv$
$[\langle$ Partially discharged proof obligations. 5.5.6.1 $\rangle$
$\vdash \langle$ Assembly of main verification. 5.5.6.5 $\rangle$
$\quad \therefore cVer_{HLA} \sqsubseteq_{Retr_{HLA}} Ver_{HLA}$
$]$

This code is used in section 5.5.

**5.5.6.1.**   Remember that the proof obligations appeared at three different places in the development. First, conditions concerning the correctness of operations: the *implementability* and *invariant preservation* obligations. Secondly, requirements about the reification correctness: the *domain* and *result* obligations. Finally, two constraints on the retrieve function: the *invariant preservation* and *completeness* obligations. All these conditions were detailed in Sect. 5.3.

$\langle$ Partially discharged proof obligations. 5.5.6.1 $\rangle \equiv$
$[\![\langle$ Proof obligations for the abstract operations. 5.5.6.2 $\rangle$
$;\langle$ Proof obligations for the concrete operations. 5.5.6.3 $\rangle$
$;\langle$ Proof obligations for the operation reifications. 5.5.6.4 $\rangle$
$;\ retrieve\_lemma\ :\ valid\_retrieve\ (cInv_{HLA}, Inv_{HLA}, Retr_{HLA})$
$]\!]$

This code is used in section 5.5.6.

**5.5.6.2.**   The proof obligations of the abstract obligations read as follows. Note that since only *DetPosNeg* has been specified, one could at most prove the second of these obligations.

⟨Proof obligations for the abstract operations. 5.5.6.2⟩ ≡

⟦ *Init_valid*          : *op_valid* (*Init*, *Inv*$_{HLA}$)

; *DetPosNeg_valid* : *op_valid* (*DetPosNeg*, *Inv*$_{HLA}$)

; *DetFail_valid*     : *op_valid* (*DetFail*, *Inv*$_{HLA}$)

; *ElimPos_valid*    : *op_valid* (*ElimPos*, *Inv*$_{HLA}$)

; *ElimNeg_valid*    : *op_valid* (*ElimNeg*, *Inv*$_{HLA}$)

; *ElimPos2_valid*  : *op_valid* (*ElimPos2*, *Inv*$_{HLA}$)

; *DetSer_valid*     : *op_valid* (*DetSer*, *Inv*$_{HLA}$)

⟧

This code is used in section 5.5.6.1.


**5.5.6.3.**  The proof obligations of the concrete obligations read as follows.

⟨Proof obligations for the concrete operations. 5.5.6.3⟩ ≡

⟦ *cInit_valid*          : *op_valid* (*cInit*, *cInv*$_{HLA}$)

; *cDetPosNeg_valid* : *op_valid* (*cDetPosNeg*, *cInv*$_{HLA}$)

; *cDetFail_valid*     : *op_valid* (*cDetFail*, *cInv*$_{HLA}$)

; *cElimPos_valid*    : *op_valid* (*cElimPos*, *cInv*$_{HLA}$)

; *cElimNeg_valid*    : *op_valid* (*cElimNeg*, *cInv*$_{HLA}$)

; *cElimPos2_valid*  : *op_valid* (*cElimPos2*, *cInv*$_{HLA}$)

; *cDetSer_valid*     : *op_valid* (*cDetSer*, *cInv*$_{HLA}$)

⟧

This code is used in section 5.5.6.1.


**5.5.6.4.**  Finally the proof obligations for the operation reification are presented. The second obligation will be proven. The presentation of the proof is deferred for now. Other obligations could be proven, after more operations have been specified.

⟨Proof obligations for the operation reifications. 5.5.6.4⟩ ≡

⟦ *Init_reif*          :   *cInit* $\sqsubseteq^{op}_{cInv_{HLA}, Retr_{HLA}}$ *Init*

; *DetPosNeg_reif* := ⟨Proof of the *DetPosNeg* operation reification. 5.5.7⟩

$$\therefore cDetPosNeg \sqsubseteq^{op}_{cInv_{HLA}, Retr_{HLA}} DetPosNeg$$

; *DetFail_reif*     :   *cDetFail* $\sqsubseteq^{op}_{cInv_{HLA}, Retr_{HLA}}$ *DetFail*

; *ElimPos_reif*    :   *cElimPos* $\sqsubseteq^{op}_{cInv_{HLA}, Retr_{HLA}}$ *ElimPos*

; *ElimNeg_reif*    :   *cElimNeg* $\sqsubseteq^{op}_{cInv_{HLA}, Retr_{HLA}}$ *ElimNeg*

; *ElimPos2_reif*    :    $cElimPos2 \sqsubseteq^{op}_{cInv_{HLA}, Retr_{HLA}} ElimPos2$

; *DetSer_reif*    :    $cDetSer \sqsubseteq^{op}_{cInv_{HLA}, Retr_{HLA}} DetSer$

⟧

This code is used in section 5.5.6.1.

**5.5.6.5.** As defined in Sect. 5.3.3, to prove that the concrete version properly reifies the abstract version, we need three proofs. These structurally somewhat complex proofs are constructed by suitably combining the verification obligations enumerated above.

⟨ Assembly of main verification. 5.5.6.5 ⟩ ≡

⦅ *version*    := ⦅ *concrete* := ⟨ Concrete version verification. 5.5.6.6 ⟩

$\therefore cVer_{HLA}$ ✓

, *abstract* := ⟨ Abstract version verification. 5.5.6.10 ⟩

$\therefore Ver_{HLA}$ ✓

⦆

, *retrieval*    := *retrieve_lemma*

$\therefore valid\_retrieve(cInv_{HLA}, Inv_{HLA}, Retr_{HLA})$

, *reification* := ⟨ Reification verification. 5.5.6.14 ⟩

$\therefore version\_reif(cVer_{HLA}, Ver_{HLA}, Retr_{HLA})$

⦆

$\therefore cVer_{HLA} \sqsubseteq_{Retr_{HLA}} Ver_{HLA}$

This code is used in section 5.5.6.

**5.5.6.6.** The first step consists in constructing the version verification proof for the concrete version. It consists of sequentially combining the proofs of the validity conditions for each operation. The presentation is split into four parts.

⟨ Concrete version verification. 5.5.6.6 ⟩ ≡

⦅ *base*    := ⦅ *base*    := ⟨ First five concrete operations. 5.5.6.7 ⟩

, *newop* := *cElimPos2_valid*

⦆

\ *def*✓ . *cons* . *up*

, *newop* := *cDetSer_valid*

⦆

\ *def*✓ . *cons* . *up*

This code is used in section 5.5.6.5.

**5.5.6.7.** ⟨ First five concrete operations. 5.5.6.7 ⟩ ≡
⟨ *base* := ⟨ *base* := ⟨ First three concrete operations. 5.5.6.8 ⟩
     , *newop* := *cElimPos_valid*
    ⟩
     \ *def*$_\checkmark$ . *cons* . *up*
, *newop* := *cElimNeg_valid*
⟩
\ *def*$_\checkmark$ . *cons* . *up*

This code is used in section 5.5.6.6.

**5.5.6.8.** ⟨ First three concrete operations. 5.5.6.8 ⟩ ≡
⟨ *base* := ⟨ *base* := ⟨ Concrete initialization. 5.5.6.9 ⟩
     , *newop* := *cDetPosNeg_valid*
    ⟩
     \ *def*$_\checkmark$ . *cons* . *up*
, *newop* := *cDetFail_valid*
⟩
\ *def*$_\checkmark$ . *cons* . *up*

This code is used in section 5.5.6.7.

**5.5.6.9.** ⟨ Concrete initialization. 5.5.6.9 ⟩ ≡
⟨ *base* := *def*$_\checkmark$.*empty* ∴ [](*inv* := *cInv*$_{HLA}$) $\checkmark$
, *newop* := *cInit_valid*
⟩
\ *def*$_\checkmark$ . *cons* . *up*

This code is used in section 5.5.6.8.

**5.5.6.10.** The proof assembly for the abstract version verification runs exactly similar.

⟨ Abstract version verification. 5.5.6.10 ⟩ ≡
⟨ *base* := ⟨ *base* := ⟨ First five abstract operations. 5.5.6.11 ⟩
     , *newop* := *ElimPos2_valid*
    ⟩
     \ *def*$_\checkmark$ . *cons* . *up*
, *newop* := *DetSer_valid*
⟩
\ *def*$_\checkmark$ . *cons* . *up*

This code is used in section 5.5.6.5.

**5.5.6.11.**    $\langle$ First five abstract operations. 5.5.6.11 $\rangle \equiv$

$\langle\!\langle$ *base*    $:= \langle\!\langle$ *base*    $:= \langle$ First three abstract operations. 5.5.6.12 $\rangle$

         , *newop* $:=$ *ElimPos_valid*

         $\rangle\!\rangle$

         $\setminus$ *def*$_\checkmark$ . *cons* . *up*

, *newop* $:=$ *ElimNeg_valid*

$\rangle\!\rangle$

$\setminus$ *def*$_\checkmark$ . *cons* . *up*

This code is used in section 5.5.6.10.

**5.5.6.12.**    $\langle$ First three abstract operations. 5.5.6.12 $\rangle \equiv$

$\langle\!\langle$ *base*    $:= \langle\!\langle$ *base*    $:= \langle$ Abstract initialization. 5.5.6.13 $\rangle$

         , *newop* $:=$ *DetPosNeg_valid*

         $\rangle\!\rangle$

         $\setminus$ *def*$_\checkmark$ . *cons* . *up*

, *newop* $:=$ *DetFail_valid*

$\rangle\!\rangle$

$\setminus$ *def*$_\checkmark$ . *cons* . *up*

This code is used in section 5.5.6.11.

**5.5.6.13.**    $\langle$ Abstract initialization. 5.5.6.13 $\rangle \equiv$

$\langle\!\langle$ *base*    $:=$ *def*$_\checkmark$ .*empty* $\therefore$ $[](inv := Inv_{HLA})$ $\checkmark$

, *newop* $:=$ *Init_valid*

$\rangle\!\rangle$

$\setminus$ *def*$_\checkmark$ . *cons* . *up*

This code is used in section 5.5.6.12.

**5.5.6.14.**    Finally, the reification verifications are assembled as follows.

$\langle$ Reification verification. 5.5.6.14 $\rangle \equiv$

$\langle\!\langle$ *base*    $:= \langle\!\langle$ *base*    $:= \langle$ First five operations. 5.5.6.15 $\rangle$

         , *newop* $:=$ *ElimPos2_reif*

         $\rangle\!\rangle$

         $\setminus$ *def_version_reif* . *cons* . *up*

, *newop* $:=$ *DetSer_reif*

$\rangle\!\rangle$

$\backslash$ *def_version_reif . cons . up*

This code is used in section 5.5.6.5.

**5.5.6.15.** $\langle$ First five operations. 5.5.6.15 $\rangle \equiv$

$\{$ *base* := $\{$ *base* := $\langle$ First three operations. 5.5.6.16 $\rangle$

  , *newop* := *ElimPos_reif*

  $\flat$

  $\backslash$ *def_version_reif . cons . up*

, *newop* := *ElimNeg_reif*

$\flat$

$\backslash$ *def_version_reif . cons . up*

This code is used in section 5.5.6.14.

**5.5.6.16.** $\langle$ First three operations. 5.5.6.16 $\rangle \equiv$

$\{$ *base* := $\{$ *base* := $\langle$ Initialization. 5.5.6.17 $\rangle$

  , *newop* := *DetPosNeg_reif*

  $\flat$

  $\backslash$ *def_version_reif . cons . up*

, *newop* := *DetFail_reif*

$\flat$

$\backslash$ *def_version_reif . cons . up*

This code is used in section 5.5.6.15.

**5.5.6.17.** $\langle$ Initialization. 5.5.6.17 $\rangle \equiv$

$\{$ *base* := *def_version_reif . empty*

  $\therefore$ *version_reif*($[](inv := cInv_{HLA})$, $[](inv := Inv_{HLA})$, $Retr_{HLA}$)

, *newop* := *Init_reif*

$\flat$

$\backslash$ *def_version_reif . cons . up*

This code is used in section 5.5.6.16.

### 5.5.7 Proof of an Operation Reification

It is interesting to observe how proofs are driven in Deva. An illustration is proposed on the operation reification obligation for the *DetPosNeg* operation, i.e., *DetPosNeg* preserves postconditions when going from the concrete state to the abstract state. This is proof *DetPosNeg_reif*, mentioned earlier but not yet presented.

The global structure of the proof matches the structure of the operation reification condition. Note that, since the precondition of *DetPosNeg* is *true*, the *domain* component of the proof becomes trivial.

⟨ Proof of the *DetPosNeg* operation reification. 5.5.7 ⟩ ≡
[ $i$ ? *void* ; $cS_i$ ? $cState_{HLA}$ ; $S_i$ := $Retr_{HLA}(cS_i)$
⊢[ *hyp_cinv* : $cInv_{HLA}(cS_i)$
  ; *hyp_pre*  : *DetPosNeg* $(i, S_i)$. *pre*
  ⊢⟨ *domain* := *true_proof* ∴ $cDetPosNeg(i, cS_i)$. *pre*
    ·, *result*    := [ $o$ ? *void* ; $cS_o$ ? $cState_{HLA}$ ; $S_o$ := $Retr_{HLA}(cS_o)$
                    ⊢[ *hyp_post* : $cDetPosNeg(i, cS_i)$. *post* $(o, cS_o)$
                      ⊢⟨ Proof of $DetPosNeg(i, S_i)$ .$post(o, S_o)$. 5.5.7.2 ⟩
                      ]
                    ]
  ⟩
]
]

This code is used in section 5.5.6.4.

**5.5.7.1.**   The *result* component consists of a proof that postcondition of the *DetPosNeg* operation holds under the given assumptions. This postcondition is a conjunction of four equations, each describing the modification of one of the components of the state. This suggests to split the proof into four subproofs.

**5.5.7.2.**    ⟨ Proof of $DetPosNeg(i, S_i)$ .$post(o, S_o)$. 5.5.7.2 ⟩ ≡
[⟨ Auxiliary Deductions. 5.5.7.3 ⟩
⊢⟨ Proof of the *Resexp* equation. 5.5.7.7 ⟩
  \ *conj*. *intro* (⟨ Proof of the *Fail* equation. 5.5.7.6 ⟩
  \ *conj*. *intro* (⟨ Proof of the *Neg* equation. 5.5.7.5 ⟩
  \ *conj*. *intro* (⟨ Proof of the *Pos* equation. 5.5.7.4 ⟩)))
]
    ∴ $DetPosNeg(i, S_i)$. *post* $(o, S_o)$

This code is used in section 5.5.7.

**5.5.7.3.**   For convenience, the assumption of the postcondition of *cDetPosNeg* is split into its conjuncts.

⟨ Auxiliary Deductions. 5.5.7.3 ⟩ ≡

⟦ hyp_Pos       := hyp_post ∖ conj . eliml ∖ conj . eliml ∖ conj . eliml
; hyp_Neg       := hyp_post ∖ conj . eliml ∖ conj . eliml ∖ conj . elimr
; hyp_Fail      := hyp_post ∖ conj . eliml ∖ conj . elimr
; hyp_Resexp    := hyp_post ∖ conj . elimr
⟧

This code is used in section 5.5.7.2.


**5.5.7.4.** The proof of the *Pos* equation is presented in a goal-oriented manner.


⟨ Proof of the *Pos* equation. 5.5.7.4 ⟩ ≡

[ goal := $Pos\,(S_o) = Cell * [\,e : Exper \vdash Res(e) = +\,] \triangleright Resexp(S_i)$
⊢[ u : goal ⊢ u ]

$$\therefore \frac{Pos\,(S_o) = Cell * [\,e : Exper \vdash Res(e) = +\,] \triangleright Resexp(S_i)}{goal}$$

⟁ $rsubst(sel_{HLA}.Resexp)$
⟁ $rsubst(sel_{HLA}.Pos)$

$$\therefore \frac{Retr_2\,(cPos(cS_o)) = Cell * [\,e : Exper \vdash Res(e) = +\,] \triangleright Retr_1(cResexp(cS_i))}{goal}$$

⟁ $rsubst(hyp\_Pos)$

$$\therefore \frac{Retr_2\,(cellfilter(+, cResexp(cS_i))) = Cell * [\,e : Exper \vdash Res(e) = +\,] \triangleright Retr_1(cResexp(cS_i))}{goal}$$

∕ $retr\_lemma_2$
    ∴ goal
]

This code is used in section 5.5.7.2.


**5.5.7.5.** The proof of the *Neg* equation is very similar to the *Pos* case. This time, no intermediate judgements are given.


⟨ Proof of the *Neg* equation. 5.5.7.5 ⟩ ≡

$[\,goal := Neg\,(S_o) = Cell * [\,e : Exper \vdash Res(e) = -\,] \rhd Resexp(S_i)$

$\vdash [\,u : goal \vdash u\,]$

$$\therefore \dfrac{Neg\,(S_o) = Cell * [\,e : Exper \vdash Res(e) = -\,] \rhd Resexp(S_i)}{goal}$$

$\oslash\ rsubst(sel_{HLA}.Resexp)$

$\oslash\ rsubst(sel_{HLA}.Neg)$

$\oslash\ rsubst(hyp\_Neg)$

$\diagup\ retr\_lemma_2$

$\qquad \therefore goal$

$]$

This code is used in section 5.5.7.2.

**5.5.7.6.**   The proof of the *Fail* equation is very easy, since both the abstract and the concrete version of *DetPosNeg* do not change this component of the state.

$\langle$ Proof of the *Fail* equation. 5.5.7.6 $\rangle \equiv$

$[\,goal := Fail\,(S_o) = Fail(S_i)$

$\vdash [\,u : goal \vdash u\,]$

$\qquad \therefore [\,Fail\,(S_o) = Fail(S_i) \vdash goal\,]$

$\oslash\ rsubst(sel_{HLA}.Fail)$

$\oslash\ rsubst(sel_{HLA}.Fail)$

$\qquad \therefore [\,Retr_2\,(cFail(cS_o)) = Retr_2(cFail(cS_i)) \vdash goal\,]$

$\oslash\ rsubst(hyp\_Fail)$

$\qquad \therefore [\,Retr_2\,(cFail(cS_i)) = Retr_2(cFail(cS_i)) \vdash goal\,]$

$\diagup\ refl$

$\qquad \therefore goal$

$]$

This code is used in section 5.5.7.2.

**5.5.7.7.**   The proof of the *Resexp* equation is very similar to the *Fail* case. Again, no intermediate judgements are given.

$\langle$ Proof of the *Resexp* equation. 5.5.7.7 $\rangle \equiv$

$[\, goal \; := \; Resexp\,(S_o) = Resexp(S_i)$

$\vdash [\, u : goal \vdash u \,]$

$\qquad \therefore [\, Resexp\,(S_o) = Resexp(S_i) \vdash goal \,]$

$\quad \diamond\; rsubst(sel_{HLA}.Resexp)$

$\quad \diamond\; rsubst(sel_{HLA}.Resexp)$

$\quad \diamond\; rsubst(hyp\_Resexp)$

$\;{}^{/}\; refl$

$\qquad \therefore goal$

$]$

This code is used in section 5.5.7.2.

### 5.5.8   Proof of another Property of the Retrieve Function

The proof of the first of the two lemmas about the retrieve function will now be presented. The basic idea is to use induction on sequences. This yields the following global structure of the proof.

$\langle\,$Proof of the first retrieve lemma. $5.5.8\,\rangle \equiv$

$[\, x \;?\; set\,(\mathcal{A}); \; l \;?\; seq\,(cExper); \; r \;?\; Result$

$$\vdash[\; thesis \; := \dfrac{l \;:\; seq\,(cExper)}{x \in Retr_2(cellfilter(r,l)) \Leftrightarrow (x \mapsto r) \in Retr_1(l)}$$

$\; ; \; base \quad := \langle\,$Proof of inductive base. $5.5.8.3\,\rangle \therefore thesis(\langle\rangle)$

$\; ; \; step \quad := [\, e \;:\; cExper \; ; \; l \;:\; seq\,(cExper); \; induct\_hyp \;:\; thesis\,(l)$

$\qquad\qquad \vdash\langle\,$Proof of inductive step. $5.5.8.4\,\rangle \therefore thesis(e :: l)$

$\qquad\qquad ]$

$\; \vdash seq\_induction\;(P := thesis)(base, step)(x := l)$

$\qquad \therefore thesis(l)$

$\quad {}^{\backslash}\; equiv\_props.\; switch\,.\, down$

$\; ]$

$]$

This code is used in section 5.5.4.5.

**5.5.8.1.**   The inductive base is proven in backward style (cf. Sec. 2.2.5) using the cut. A highly implicit version of the proof makes use of the evaluation tactic of the HLA case study.

**5.5.8.2.**   $\langle\,$Proof of inductive base (not checked). $5.5.8.2\,\rangle \equiv$

$[\, u : thesis\,(\langle\rangle) \vdash u \,]$

◇ **loop** *HLAEvaluation . empty*

∴ $[true \vdash thesis(\langle\rangle)]$

/ *true_proof*

**5.5.8.3.** By unfolding the evaluation tactic, one obtains a stepwise proof of the inductive base.

⟨ Proof of inductive base. 5.5.8.3 ⟩ ≡

$[u : thesis(\langle\rangle) \vdash u]$

$$\therefore \left|\begin{array}{l} x \in Retr_2(cCell *_{seq} ([e : cExper \vdash cRes(e) = r] \triangleright_{seq} \langle\rangle)) \\ \quad \Leftrightarrow (x \mapsto r) \in Retr_1(\langle\rangle) \\ \hline \qquad\qquad thesis(\langle\rangle) \end{array}\right.$$

◇ $rsubst(def_{\triangleright_{seq}}.base)$

∴ $[x \in Retr_2(cCell *_{seq} \langle\rangle) \Leftrightarrow (x \mapsto r) \in Retr_1(\langle\rangle) \vdash thesis(\langle\rangle)]$

◇ $rsubst(def_{*_{seq}}.base)$

∴ $[x \in Retr_2(\langle\rangle) \Leftrightarrow (x \mapsto r) \in Retr_1(\langle\rangle) \vdash thesis(\langle\rangle)]$

◇ $rsubst(Retr_2\_def.base)$

∴ $[x \in \{\} \Leftrightarrow (x \mapsto r) \in Retr_1(\langle\rangle) \vdash thesis(\langle\rangle)]$

◇ $rsubst(Retr_1\_def.base)$

∴ $[x \in \{\} \Leftrightarrow (x \mapsto r) \in \{\} \vdash thesis(\langle\rangle)]$

◇ $prsubst(equiv\_props. inj\_fl (def_{\in}.base))$

∴ $[false \Leftrightarrow (x \mapsto r) \in \{\} \vdash thesis(\langle\rangle)]$

◇ $prsubst(equiv\_props. inj\_fl (def_{\in}.base))$

∴ $[false \Leftrightarrow false \vdash thesis(\langle\rangle)]$

◇ $equiv\_props. refl . up$

∴ $[true \vdash thesis(\langle\rangle)]$

/ *true_proof*

This code is used in section 5.5.8.

**5.5.8.4.** To prove the inductive step, one has to realize that a case distinction must be made of whether the reaction with the new experiment is positive or negative. This leads to the following breakdown of the proof.

⟨ Proof of inductive step. 5.5.8.4 ⟩ ≡

$[\ step\_eq\ \ :=[\ eq\_hyp\ :\ cRes\,(e)=r$

$\qquad\qquad \vdash \langle$ Case of positive reaction. $5.5.8.6\,\rangle\ \therefore\ thesis(e::l)$

$\qquad\qquad ]$

$;\ step\_neq\ :=[\ neq\_hyp\ :\ \neg\ cRes\,(e)=r$

$\qquad\qquad \vdash \langle$ Case of negative reaction. $5.5.8.7\,\rangle\ \therefore\ thesis(e::l)$

$\qquad\qquad ]$

$\vdash disj\ .\ elim\ (excluded\_middle, step\_eq, step\_neq)$

$]$

This code is used in section 5.5.8.

**5.5.8.5.** $\quad\langle$ Case of positive reaction (not checked). $5.5.8.5\,\rangle\equiv$

$[\ u: thesis\ (e::l)\vdash u\,]$

$\quad\therefore\ [\ thesis\ (e::l)\vdash thesis(e::l)\,]$

$\qquad\begin{array}{|l}x\in Retr_2(cCell *_{seq}\,([\ e1:cExper\ \vdash cRes(e1)=r\,]\ \triangleright_{seq}\ (e::l)))\\[4pt]\quad\Leftrightarrow (x\mapsto r)\in Retr_1(e::l)\\ \hline \qquad\qquad thesis\ (e::l)\end{array}$

$\therefore$

$\diamond\ rsubst(def_{\triangleright_{seq}}(h:=e).\ rec\ .\mathbf{1}(P:=[\ e1:cExper\ \vdash cRes(e1)=r\,],\ eq\_hyp))$

$\diamond\ \mathbf{loop}\ HLA\,Evaluation\ .\ cons$

$\qquad\begin{array}{|l}(x=elems(cCell(e)))\vee(x\in Retr_2(cellfilter(r,l)))\\[4pt]\quad\Leftrightarrow (x=elems(cCell(e))\wedge r=cRes(e))\vee(x\mapsto r)\in Retr_1(l)\\ \hline \qquad\qquad thesis\ (e::l)\end{array}$

$\therefore$

$\diamond\ psubst(equiv\_props.\ switch\ .\ down\ (induct\_hyp))$

$\qquad\begin{array}{|l}(x=elems(cCell(e)))\vee(x\in Retr_2(cellfilter(r,l)))\\[4pt]\quad\Leftrightarrow (x=elems(cCell(e))\wedge r=cRes(e))\vee(x\in Retr_2(cellfilter(r,l)))\\ \hline \qquad\qquad thesis\ (e::l)\end{array}$

$\therefore$

$\diamond\ equiv\_props.\ mon\_orl\ .\ down$

$\qquad\begin{array}{|l}x=elems(cCell(e))\Leftrightarrow(x=elems(cCell(e))\wedge r=cRes(e))\\ \hline \qquad\qquad thesis\ (e::l)\end{array}$

$\therefore$

$\diamond\ equiv\_props.\ conj\_right\ (sym(eq\_hyp)).\ down$

$\qquad\begin{array}{|l}x=elems(cCell(e))\Leftrightarrow x=elems(cCell(e))\\ \hline \qquad\qquad thesis\ (e::l)\end{array}$

$\therefore$

$\diamond\ equiv\_props.\ refl\ .\ up$

$\qquad\begin{array}{|l}true\\ \hline thesis\ (e::l)\end{array}$

$\therefore$

$/\ true\_proof$

**5.5.8.6.** The proof for the case of a positive reaction reads as follows:

$\langle$ Case of positive reaction. $5.5.8.6\,\rangle\equiv$

$[\,u : thesis\ (e :: l) \vdash u\,]$

$\therefore\ [\,thesis\ (e :: l) \vdash thesis(e :: l)\,]$

$\therefore\ \left|\ \begin{array}{l} x \in Retr_2(cCell *_{seq} (\,[\,e1 : cExper \vdash cRes(e1) = r\,] \rhd_{seq} (e :: l))) \\ \quad \Leftrightarrow (x \mapsto r) \in Retr_1(e :: l) \\ \hline \hphantom{xxxxxxxxxx} thesis\ (e :: l) \end{array}\right.$

$\Diamond\ rsubst(def_{\rhd_{seq}}.\ rec\ .\mathbf{1}(P := [\,e1 : cExper \vdash cRes(e1) = r\,], eq\_hyp))$

$\therefore\ \left|\ \begin{array}{l} x \in Retr_2(cCell *_{seq} (e :: (\,[\,e1 : cExper \vdash cRes(e1) = r\,] \rhd_{seq} l))) \\ \quad \Leftrightarrow (x \mapsto r) \in Retr_1(e :: l) \\ \hline \hphantom{xxxxxxxxxx} thesis\ (e :: l) \end{array}\right.$

$\Diamond\ rsubst(def_{*_{seq}}.rec)$

$\therefore\ \left|\ \begin{array}{l} x \in Retr_2(cCell(e) :: cellfilter(r, l)) \Leftrightarrow (x \mapsto r) \in Retr_1(e :: l) \\ \hline \hphantom{xxxxxxxxxx} thesis\ (e :: l) \end{array}\right.$

$\Diamond\ rsubst(Retr_2\_def.rec)$

$\therefore\ \left|\ \begin{array}{l} x \in elems(cCell(e)) \odot Retr_2(cellfilter(r, l)) \Leftrightarrow (x \mapsto r) \in Retr_1(e :: l) \\ \hline \hphantom{xxxxxxxxxx} thesis\ (e :: l) \end{array}\right.$

$\Diamond\ rsubst(Retr_1\_def.rec)$

$\therefore\ \left|\ \begin{array}{l} x \in elems(cCell(e)) \odot Retr_2(cellfilter(r, l)) \\ \quad \Leftrightarrow (x \mapsto r) \in (elems(cCell(e)) \mapsto cRes(e)) \odot Retr_1(l) \\ \hline \hphantom{xxxxxxxxxx} thesis\ (e :: l) \end{array}\right.$

$\Diamond\ prsubst(def_{\in}.rec)$

$\therefore\ \left|\ \begin{array}{l} (x = elems(cCell(e))) \lor (x \in Retr_2(cellfilter(r, l))) \\ \quad \Leftrightarrow (x \mapsto r) \in (elems(cCell(e)) \mapsto cRes(e)) \odot Retr_1(l) \\ \hline \hphantom{xxxxxxxxxx} thesis\ (e :: l) \end{array}\right.$

$\Diamond\ prsubst(def_{\in}.rec)$

$\therefore\ \left|\ \begin{array}{l} (x = elems(cCell(e))) \lor (x \in Retr_2(cellfilter(r, l))) \\ \quad \Leftrightarrow (x \mapsto r) = (elems(cCell(e)) \mapsto cRes(e)) \lor (x \mapsto r) \in Retr_1(l) \\ \hline \hphantom{xxxxxxxxxx} thesis\ (e :: l) \end{array}\right.$

$\Diamond\ prsubst(mkbin\_injective)$

$\therefore\ \left|\ \begin{array}{l} (x = elems(cCell(e))) \lor (x \in Retr_2(cellfilter(r, l))) \\ \quad \Leftrightarrow (x = elems(cCell(e)) \land r = cRes(e)) \lor (x \mapsto r) \in Retr_1(l) \\ \hline \hphantom{xxxxxxxxxx} thesis\ (e :: l) \end{array}\right.$

$\Diamond\ psubst(equiv\_props.\ switch\ .\ down\ (induct\_hyp))$

$\therefore\ \left|\ \begin{array}{l} (x = elems(cCell(e))) \lor (x \in Retr_2(cellfilter(r, l))) \\ \quad \Leftrightarrow (x = elems(cCell(e)) \land r = cRes(e)) \lor (x \in Retr_2(cellfilter(r, l))) \\ \hline \hphantom{xxxxxxxxxx} thesis\ (e :: l) \end{array}\right.$

$\Diamond\ equiv\_props.\ mon\_orl\ .\ down$

$\therefore\ [\,x = elems(cCell(e)) \Leftrightarrow (x = elems(cCell(e)) \land r = cRes(e)) \vdash thesis(e :: l)\,]$

$\Diamond\ equiv\_props.\ conj\_right\ (sym(eq\_hyp)).\ down$

$\therefore [x = elems(cCell(e)) \Leftrightarrow x = elems(cCell(e)) \vdash thesis(e :: l)]$

$\diamond equiv\_props.\ refl\ .\ up$

$\quad \therefore [true \vdash thesis(e :: l)]$

$/\ true\_proof$

This code is used in section 5.5.8.4.

**5.5.8.7.** The proof for the case of a negative reaction reads as follows:

$\langle$ Case of negative reaction. 5.5.8.7 $\rangle \equiv$

$[u : thesis\ (e :: l) \vdash u]$

$\quad \therefore [thesis\ (e :: l) \vdash thesis(e :: l)]$

$$\therefore \frac{\begin{array}{l} x \in Retr_2(cCell *_{seq} ([el : cExper \vdash cRes(el) = r] \rhd_{seq} (e :: l))) \\ \Leftrightarrow (x \mapsto r) \in Retr_1(e :: l) \end{array}}{thesis\ (e :: l)}$$

$\diamond rsubst(def_{\rhd_{seq}}.\ rec\ .2(P := [el : cExper \vdash cRes(el) = r], neq\_hyp))$

$$\therefore \frac{\begin{array}{l} x \in Retr_2(cCell *_{seq} ([el : cExper \vdash cRes(el) = r] \rhd_{seq} l)) \\ \Leftrightarrow (x \mapsto r) \in Retr_1(e :: l) \end{array}}{thesis\ (e :: l)}$$

$$\therefore \frac{x \in Retr_2(cellfilter(r, l)) \Leftrightarrow (x \mapsto r) \in Retr_1(e :: l)}{thesis\ (e :: l)}$$

$\diamond prsubst(equiv\_props.\ switch\ .\ down\ (induct\_hyp))$

$$\therefore \frac{(x \mapsto r) \in Retr_1(l) \Leftrightarrow (x \mapsto r) \in Retr_1(e :: l)}{thesis\ (e :: l)}$$

$\diamond rsubst(Retr_1\_def.rec)$

$$\therefore \frac{(x \mapsto r) \in Retr_1(l) \Leftrightarrow (x \mapsto r) \in (elems(cCell(e)) \mapsto cRes(e)) \odot Retr_1(l)}{thesis\ (e :: l)}$$

$\diamond prsubst(def_{\in}.rec)$

$$\therefore \frac{\begin{array}{l} (x \mapsto r) \in Retr_1(l) \\ \Leftrightarrow (x \mapsto r) = (elems(cCell(e)) \mapsto cRes(e)) \vee (x \mapsto r) \in Retr_1(l) \end{array}}{thesis\ (e :: l)}$$

$\diamond prsubst(mkbin\_injective)$

$$\therefore \frac{\begin{array}{l} (x \mapsto r) \in Retr_1(l) \\ \Leftrightarrow (x = elems(cCell(e)) \wedge r = cRes(e)) \vee (x \mapsto r) \in Retr_1(l) \end{array}}{thesis\ (e :: l)}$$

$\diamond prsubst(equiv\_props.\ inj\_fl\ (symneg(neq\_hyp)))$

$$\therefore \frac{(x \mapsto r) \in Retr_1(l) \Leftrightarrow (x = elems(cCell(e)) \wedge false) \vee (x \mapsto r) \in Retr_1(l)}{thesis\ (e :: l)}$$

$\diamond prsubst(conj\_props.\ simpr\ .false)$

$$\therefore \left| \frac{(x \mapsto r) \in Retr_1(l) \Leftrightarrow false \vee (x \mapsto r) \in Retr_1(l)}{thesis\,(e :: l)} \right.$$

$\diamondsuit\ prsubst(disj\_props.\ simpl\ .false)$

$$\therefore \left| \frac{(x \mapsto r) \in Retr_1(l) \Leftrightarrow (x \mapsto r) \in Retr_1(l)}{thesis\,(e :: l)} \right.$$

$\diamondsuit\ equiv\_props.\ refl\ .\ up$

$\therefore [true \vdash thesis(e :: l)]$

$/\ true\_proof$

This code is used in section 5.5.8.4.

## 5.6  Proof of Transitivity of Reification in Deva

This section presents a proof that reification, as defined in the previous section, is transitive. More precisely, for arbitrary versions $v_1, v_2, v_3$ and retrieve functions $retr_{12}$, $retr_{23}$, the following property is proven:

$$\left| \frac{v_1 \sqsubseteq_{retr_{12}} v_2;\ v_2 \sqsubseteq_{retr_{23}} v_3}{v_1 \sqsubseteq_{retr_{12} \diamondsuit retr_{23}} v_3} \right.$$

Figuratively speaking the above property states that successive reifications can be composed by composing their retrieve functions.

### 5.6.1  Frequently Used Contexts

In order to formally state and prove the above property without long and boring object declarations, it is useful to systematically abbreviate a number of frequently used contexts introducing implicitly defined identifiers. These contexts describe situations in which (combinations of) triples of operations, retrieve functions, invariants, and versions are considered.

$\langle$ Frequently used contexts. 5.6.1 $\rangle \equiv$

**import context** $FrequentContexts :=$

$[\![\,\langle$ Invariants. 5.6.1.1 $\rangle$

$;\langle$ Operations. 5.6.1.2 $\rangle$

$;\langle$ Versions. 5.6.1.3 $\rangle$

$;\langle$ Retrieve functions. 5.6.1.4 $\rangle$

$;\langle$ Versions and operations. 5.6.1.5 $\rangle$

$;\langle$ Versions and retrieve functions. 5.6.1.6 $\rangle$

$]\!]$

This code is used in section 5.6.3.2.

**5.6.1.1.**    ⟨ Invariants. 5.6.1.1 ⟩ ≡

**context** *Invariants* :=

⟦ *state*$_1$, *state*$_2$, *state*$_3$ ? *sort*

; *inv*$_1$                    ? [ *state*$_1$ ⊢ *prop* ]

; *inv*$_2$                    ? [ *state*$_2$ ⊢ *prop* ]

; *inv*$_3$                    ? [ *state*$_3$ ⊢ *prop* ]

⟧

This code is used in section 5.6.1.

**5.6.1.2.**    ⟨ Operations. 5.6.1.2 ⟩ ≡

**context** *Operations* :=

⟦ *state*$_1$, *state*$_2$, *state*$_3$ ? *sort*

; *in*, *out*                   ? *sort*

; *op*$_1$                     ? *OP* (*in*, *out*, *state*$_1$)

; *op*$_2$                     ? *OP* (*in*, *out*, *state*$_2$)

; *op*$_3$                     ? *OP* (*in*, *out*, *state*$_3$)

⟧

This code is used in section 5.6.1.

**5.6.1.3.**    ⟨ Versions. 5.6.1.3 ⟩ ≡

**context** *Versions* :=

⟦ **import** *Invariants*

; *v*$_1$ ? *version* (*inv*$_1$)

; *v*$_2$ ? *version* (*inv*$_2$)

; *v*$_3$ ? *version* (*inv*$_3$)

⟧

This code is used in section 5.6.1.

**5.6.1.4.**    ⟨ Retrieve functions. 5.6.1.4 ⟩ ≡

**context** *RetrieveFunctions* :=

⟦ *state*$_1$, *state*$_2$, *state*$_3$ ?   *sort*

; *retr*$_{12}$                   ? [ *state*$_1$ ⊢ *state*$_2$ ]

; *retr*$_{23}$                   ? [ *state*$_2$ ⊢ *state*$_3$ ]

; *retr*$_{13}$                   := [ *st*$_1$ : *state*$_1$ ⊢ *retr*$_{23}$(*retr*$_{12}$(*st*$_1$)) ]

⟧

This code is used in section 5.6.1.

**5.6.1.5.**   ⟨ Versions and operations. 5.6.1.5 ⟩ ≡
**context** *VersionsAndOperations* :=
⟦ *state₁*, *state₂*, *state₃* ? *sort*
; **import** *Versions*
; **import** *Operations*
⟧

This code is used in section 5.6.1.

**5.6.1.6.**   ⟨ Versions and retrieve functions. 5.6.1.6 ⟩ ≡
**context** *VersionsAndRetrieveFunctions* :=
⟦ *state₁*, *state₂*, *state₃* ? *sort*
; **import** *Versions*
; **import** *RetrieveFunctions*
⟧

This code is used in section 5.6.1.

### 5.6.2   Simultaneous Induction on Version Triples

**5.6.2.1.**   In order to lift properties of operation triples to properties of version triples, an induction principle of simultaneous induction on version triples is needed. This principle can be derived using structural induction on versions (Sect. 5.3.2).

⟨ Induction principle. 5.6.2.1 ⟩ ≡
*version_triple_induction* :
⟦ **import** *Invariants*
; $P$ : ⟦ *version* ($inv_1$); *version*($inv_2$); *version*($inv_3$) ⊢ *prop* ⟧
   *base* : $P$ ($[]$, $[]$, $[]$);
   *step* : ⟦ **import** *VersionsAndOperations*

$$\dfrac{P\,(v_1,\,v_2,\,v_3)}{P\,(op_1 \circledcirc v_1,\ op_2 \circledcirc v_2,\ op_3 \circledcirc v_3)}$$

    ⟧

⊢   ⟦ **import** *Versions*

$$\dfrac{vlength\,(v_1) = vlength(v_2);\ vlength(v_2) = vlength(v_3)}{P\,(v_1,\,v_2,\,v_3)}$$

⟧
⟧

This code is used in section 5.6.2.4.

The inductive derivation of this inductive principle is omitted. Note, that the induction principles is parametrized over invariant triples and achieves induction over version triples. It is required that all three versions are of equal length.

**5.6.2.2.** The *length vlength* $(v)$ of a version $v$ is defined as the number of operations in $v$.

⟨ Length of versions. 5.6.2.2 ⟩ ≡
⟦ *vlength*      : [ *state* ? *sort* ; *inv* ? [*state* ⊢ *prop*] ⊢ [ *version* (*inv*) ⊢ *nat* ]]
; *def_length* :
  ⦃ *empty* := *vlength* ([]) = 0
  , *cons*  := [ *in, out, state* ? *sort* ;                    *inv* ? [*state* ⊢ *prop*]
              ; *op*            ? *OP* (*in, out, state*); *v*    ? *version* (*inv*)
              ⊢ *vlength* (*op* ⊙ *v*) = 1 + *vlength*(*v*)
              ]
  ⦄
⟧

This code is used in section 5.6.2.4.

**5.6.2.3.** It is easy to see that two given versions which satisfy the version reification condition are of equal length.

⟨ Reified versions are of equal length. 5.6.2.3 ⟩ ≡
*equal_length* : [ *cstate, astate* ? *sort*
              ; *cinv*            ? [*cstate* ⊢ *prop*]; *ainv* ? [*astate* ⊢ *prop*]
              ; *cv*              ? *version* (*cinv*); *av*    ? *version* (*ainv*)
              ; *retr*            ? [*cstate* ⊢ *astate*]
              ⊢ $\dfrac{version\_reif(cv, av, retr)}{vlength\,(cv) = vlength(av)}$
              ]

This code is used in section 5.6.2.4.

**5.6.2.4.** The given pieces can be assembled into a context formalizing induction on version triples.

⟨ Induction on version triples. 5.6.2.4 ⟩ ≡

$[\![\,\langle$ Length of versions. 5.6.2.2 $\rangle$

$;\langle$ Reified versions are of equal length. 5.6.2.3 $\rangle$

$;\langle$ Induction principle. 5.6.2.1 $\rangle$

$]\!]$

This code is used in section 5.6.3.2.

### 5.6.3   Global Proof Scheme

**5.6.3.1.**   The proof of the transitivity property will be constructed from the frequently used contexts and the induction principle for version triples. In the Deva calculus, this means that, using the context *FrequentContexts* and the context *InductionOnVersionTriples*, a text *refine* can be constructed which proves reification transitivity, i.e.,

**5.6.3.2.**   $\langle$ Transitivity of reification. 5.6.3.2 $\rangle \equiv$

**context**   *TransitivityofReification* :=

$[\![\,\langle$ Frequently used contexts. 5.6.1 $\rangle$

$;\langle$ Induction on version triples. 5.6.2.4 $\rangle$

$;$ *refine* := $\langle$ Proof of transitivity. 5.6.3.3 $\rangle$

$]\!]$

This code is used in section 5.1.

**5.6.3.3.**   The structure of *refine* is very analogous to that of its type in that it also assumes the premises of transitivity of reification, that is, a version $v_1$ is assumed to be a reification of a version $v_2$ wrt. the retrieve function $retr_{12}$, and a version $v_2$ is assumed to be a reification of a version $v_3$ wrt. a retrieve function $retr_{23}$:

$\langle$ Proof of transitivity. 5.6.3.3 $\rangle \equiv$

$[\,$**import**   *VersionsAndRetrieveFunctions*

$\vdash \dfrac{reif_{12} \,:\, v_1 \sqsubseteq_{retr_{12}} v_2;\; reif_{23} \,:\, v_2 \sqsubseteq_{retr_{23}} v_3}{\langle\text{Proof of } v_1 \sqsubseteq_{retr_{13}} v_3.\ 5.6.3.4\,\rangle \,\therefore\, v_1 \sqsubseteq_{retr_{13}} v_3}$

$]$

This code is used in section 5.6.3.2.

**5.6.3.4.**   The proof that $v_1$ is a reification of a version $v_3$ wrt. the retrieve function $retr_{12} \Diamond retr_{23}$ is decomposed into proofs of three lemmas:

$\langle$ Proof of $v_1 \sqsubseteq_{retr_{13}} v_3.$ 5.6.3.4 $\rangle \equiv$

[$\langle$ Verification of the retrieve condition. 5.6.4.1 $\rangle$

;$\langle$ Transitivity of operator reification. 5.6.5.1 $\rangle$

;$\langle$ Transitivity of the reification condition. 5.6.6.1 $\rangle$

$\vdash\langle$ Proof assembly. 5.6.7 $\rangle \therefore v_1 \sqsubseteq_{retr_{13}} v_3$

]

This code is used in section 5.6.3.3.

### 5.6.4 Verification of the Retrieve Condition

**5.6.4.1.** The first proposition to be shown is that the composed retrieve function $retr_{13}$ satisfies the retrieve condition.

- The property of invariant preservation *preservation* is derived by the logical cut of the *preservation* components of the two assumptions $reif_{12}$ and $reif_{23}$.
- The proof of the completeness condition *completeness* is not yet detailed at this point.

$\langle$ Verification of the retrieve condition. 5.6.4.1 $\rangle \equiv$

$retrieval_{13} :=$

$\langle$ *preservation* $:= reif_{12} \cdot retrieval \cdot preservation$

$\qquad \therefore [\, st_1\ ?\ state_1\ \vdash [\, inv_1\ (st_1) \vdash inv_2(retr_{12}(st_1))]\,]$

$\qquad \diamondsuit reif_{23}. retrieval . preservation$

$\qquad \therefore [\, st_1\ ?\ state_1\ \vdash [\, inv_1\ (st_1) \vdash inv_3(retr_{13}(st_1))]\,]$

, *completeness* $:= \langle$ Proof of completeness condition. 5.6.4.2 $\rangle$

$\rangle$

$\qquad \therefore valid\_retrieve(inv_1, inv_3, retr_{13})$

This code is used in section 5.6.3.4.

**5.6.4.2.** The proof of the completeness condition consists of showing that a given abstract state $st_3 : state_3$ that satisfies the abstract invariant is in the range of the retrieve function $retr_{13}$ from $state_1$ to $state_3$, where $state_1$ is further restricted by the concrete invariant $inv_1$ $(st_1)$.

The proof idea is to make a block-structured proof using the two completeness assumptions contained in the reification assumptions $reif_{12}$ and $reif_{23}$. The proof is given in a top-down presentation involving three parts. Unfortunately, it looks technically rather heavy; this is mainly due to the elimination law of existential quantification. The reader is advised to study again this law ( $ex$ .$elim$, see section 4.2.1.7), to understand the scoping structure of the proof.

⟨ Proof of completeness condition. 5.6.4.2 ⟩ ≡
[ $st_3$ ? $state_3$ ; $inv\_hyp$ : $inv_3$ ($st_3$)
⊢ $reif_{23}$ . $retrieval$ . $completeness$ ($inv\_hyp$)
$\quad\quad ∴ ∃[ st_2 : state_2 ⊢ retr_{23}(st_2) = st_3 ∧ inv_2(st_2) ]$

$\quad\quad\quad\quad\quad$ | $st_2$ : $state_2$ ; $hyp$ : $retr_{23}$ ($st_2$) = $st_3$ ∧ $inv_2$($st_2$)
$\quad$ \ $ex. elim$ ( |⟨First block. 5.6.4.3 ⟩$\quad\quad\quad\quad\quad\quad\quad\quad\quad\quad\quad\quad$ )
$\quad\quad\quad\quad\quad$ | $\quad ∴ ∃[ st_1 : state_1 ⊢ retr_{13}(st_1) = st_3 ∧ inv_1(st_1) ]$
$\quad\quad ∴ ∃[ st_1 : state_1 ⊢ retr_{13}(st_1) = st_3 ∧ inv_1(st_1) ]$
]

This code is used in section 5.6.4.1.

**5.6.4.3.**   ⟨First block. 5.6.4.3 ⟩ ≡
$reif_{12}$ . $retrieval$ . $completeness$ ($conj. elimr$ ($hyp$))
$\quad\quad ∴ ∃[ st_1 : state_1 ⊢ retr_{12}(st_1) = st_2 ∧ inv_1(st_1) ]$

$\quad\quad\quad\quad\quad$ | $st_1$ : $state_1$ ; $hyp1$ : $retr_{12}$ ($st_1$) = $st_2$ ∧ $inv_1$($st_1$)
$\quad$ \ $ex. elim$ ( |⟨Second block. 5.6.4.4 ⟩$\quad\quad\quad\quad\quad\quad\quad\quad\quad\quad$ )
$\quad\quad\quad\quad\quad$ | $\quad ∴ ∃[ st_1 : state_1 ⊢ retr_{13}(st_1) = st_3 ∧ inv_1(st_1) ]$
$\quad\quad ∴ ∃[ st_1 : state_1 ⊢ retr_{13}(st_1) = st_3 ∧ inv_1(st_1) ]$

This code is used in section 5.6.4.2.

**5.6.4.4.**   Given $hyp$ and $hyp1$, we now use a simple law of equational reasoning ($eq\_compose$, see section 4.2.2.6) to establish that $st_3$ is in the range of $retr_{13}$.

⟨Second block. 5.6.4.4 ⟩ ≡
$conj$ . $intro$
$\quad$ / ($eq\_compose(conj. eliml$ ($hyp1$), $conj. eliml$ ($hyp$)) ∴ $retr_{13}(st_1) = st_3$)
$\quad$ / ($conj. elimr$ ($hyp1$) ∴ $inv_1(st_1)$)
$\quad\quad ∴ retr_{13}(st_1) = st_3 ∧ inv_1(st_1)$
$\quad$ \ $ex. intro$
$\quad\quad ∴ ∃[ st_1 : state_1 ⊢ retr_{13}(st_1) = st_3 ∧ inv_1(st_1) ]$

This code is used in section 5.6.4.3.

## 5.6.5   Transitivity of Operator Reification

**5.6.5.1.**   The second property needed is a lemma stating transitivity of the operator reification condition. Assuming this condition to hold for operator pairs $op_1, op_2$ and $op_2, op_3$ respectively, the condition must be shown for the pair $op_1, op_3$:

⟨ Transitivity of operator reification. 5.6.5.1 ⟩ ≡
$op\_refine$ := [ **import** *Operations*

$$\vdash \begin{array}{|l}
op\_reif_{12} : op_1 \sqsubseteq^{op}_{inv_1, retr_{12}} op_2; \\
op\_reif_{23} : op_2 \sqsubseteq^{op}_{inv_2, retr_{23}} op_3 \\
\hline
⟨ \text{Proof of } op_1 \sqsubseteq^{op}_{inv_1, retr_{13}} op_3. \; 5.6.5.2 ⟩
\end{array}$$

]

This code is used in section 5.6.3.4.

**5.6.5.2.** A proof of the latter proposition is presented below: Its assumptions have the same structure as that of the definition of operation reification, i.e., it is assumed that the concrete invariant ($inv_1\_hyp$) and that the precondition of the abstract operation is satisfied ($op_3\_pre\_hyp$). In order to make use of the assumptions about operation reification ($op\_reif_{12}, op\_reif_{23}$), it is necessary to first derive the intermediate invariant ($inv_2\_proof$) and the precondition of the intermediate operation ($op_2\_pre\_proof$). Note, that the first proof makes use of the overall hypothesis $reif_{12}$. More precisely, it needs the assumption that $retr_{12}$ preserves the invariants. In the proof body, the fact that the precondition is preserved ($domain$) follows from the corresponding fact contained in the assumption $op\_reif_{12}$.

⟨ Proof of $op_1 \sqsubseteq^{op}_{inv_1, retr_{13}} op_3$. 5.6.5.2 ⟩ ≡
[ $i$ ? *in* ; $st_1\_i$ ? $state_1$
⊦[ $st_2\_i$ := $retr_{12} (st_1\_i)$; $st_3\_i$ := $retr_{23} (st_2\_i)$
  ⊦[ $inv_1\_hyp$ : $inv_1 (st_1\_i)$; $op_3\_pre\_hyp$ : $op_3 (i, st_3\_i)$. *pre*
    ⊦[ $inv_2\_proof$     := $reif_{12}$ . *retrieval* . *preservation* ($inv_1\_hyp$)
                        ∴ $inv_2(st_2\_i)$
    ; $op_2\_pre\_proof$ := $op\_reif_{23} (inv_2\_proof, op_3\_pre\_hyp)$. *domain*
                        ∴ $op_2(i, st_2\_i)$. *pre*
    ⊦⟨ $domain$ := $op\_reif_{12} (inv_1\_hyp, op_2\_pre\_proof)$. *domain*
                   ∴ $op_1(i, st_1\_i)$. *pre*
    , $result$    := ⟨ Proof of the $result$ case. 5.6.5.3 ⟩
      ⟩
    ]
  ]
]
]

$$\therefore op_1 \sqsubseteq^{op}_{inv_1,retr_{13}} op_3$$

This code is used in section 5.6.5.1.

**5.6.5.3.** The fact that the postcondition is preserved (*result*) can be shown by cutting the corresponding facts contained in the two assumptions $op\_reif_{12}$ and $op\_reif_{23}$.

⟨ Proof of the *result* case. 5.6.5.3 ⟩ ≡

[ $o$ ? $out$ ; $st_1\_o$ ? $state_1$

⊢[ $st_2\_o := retr_{12} (st_1\_o)$; $st_3\_o := retr_{23} (st_2\_o)$

   ⊢ $op\_reif_{12} (inv_1\_hyp, op_2\_pre\_proof)$. $result$

      $\therefore$ [ $op_1 (i, st_1\_i)$. $post (o, st_1\_o) \vdash op_2(i, st_2\_i)$. $post (o, st_2\_o)$ ]

   ⟡ $op\_reif_{23}(inv_2\_proof, op_3\_pre\_hyp)$. $result$

      $\therefore$ [ $op_1 (i, st_1\_i)$. $post (o, st_1\_o) \vdash op_3(i, st_3\_i)$. $post (o, st_3\_o)$ ]

]

]

This code is used in section 5.6.5.2.

## 5.6.6   Transitivity of the Reification Condition

**5.6.6.1.**   To lift transitivity of operator reification to the level of transitivity of the version reification condition, an inductive proof will be used.

⟨ Transitivity of the reification condition. 5.6.6.1 ⟩ ≡

[ $trans\_version\_reifctn :=$

  [ $v_1 : version (inv_1)$; $v_2 : version (inv_2)$; $v_3 : version (inv_3)$

⊢$version\_reif(v_1, v_2, retr_{12})$

    ⇒ $version\_reif(v_2, v_3, retr_{23})$

    ⇒ $version\_reif(v_1, v_3, retr_{13})$

  ]

;⟨Inductive proof of the transitivity condition. 5.6.6.2 ⟩

]

This code is used in section 5.6.3.4.

**5.6.6.2.**   The inductive proof makes use of the induction principle for version triples, instantiated to the property of transitivity of version reification condition. The inductive base and the inductive step are given spearately. At the end, after the induction principle has been applied, the conditions requiring equal length are resolved.

⟨ Inductive proof of the transitivity condition. 5.6.6.2 ⟩ ≡

⟦ *trans_base*                       := ⟨ Base case. 5.6.6.3 ⟩

; *trans_step*                       := ⟨ Inductive step. 5.6.6.4 ⟩

; *trans_version_reifctn_proof* :=

  *version_triple_induction* ($P$ := *trans_version_reifctn*,

$$\qquad\qquad\qquad base := trans\_base, step := trans\_step)$$

  ∴ ⌈ **import**   *Versions*

$$\vdash \frac{vlength\,(v_1) = vlength(v_2);\; vlength(v_2) = vlength(v_3)}{trans\_version\_reifctn\,(v_1, v_2, v_3)}$$

  ⌉

 ╱ *equal_length*(*reif*$_{12}$.*reification*)

 ╱ *equal_length*(*reif*$_{23}$.*reification*)

   ∴ *trans_version_reifctn*($v_1, v_2, v_3$)

⟧

This code is used in section 5.6.6.1.

**5.6.6.3.**   For the base case of empty versions, it essentially suffices to use the case of empty versions in the definition of the reification condition.

⟨ Base case. 5.6.6.3 ⟩ ≡

[*version_reif*(*cinv* := *inv*$_1$, *ainv* := *inv*$_2$)([], [], *retr*$_{12}$)

⊢[*version_reif*(*cinv* := *inv*$_2$, *ainv* := *inv*$_3$)([], [], *retr*$_{23}$)

  ⊢ *def_version_reif* (*cinv* := *inv*$_1$, *ainv* := *inv*$_3$). *empty*

       ∴ *version_reif*(*cinv* := *inv*$_1$, *ainv* := *inv*$_3$)([], [], *retr*$_{13}$)

  ]

   ╲ *imp. intro*

]

 ╲ *imp. intro*

   ∴ *trans_version_reifctn*([], [], [])

This code is used in section 5.6.6.2.

**5.6.6.4.**   For the inductive step, it is necessary to prove transitivity of the "consed" versions under the inductive assumption of transitivity of the original versions (*inductive_hyp*).

⟨ Inductive step. 5.6.6.4 ⟩ ≡

$[$ **import**    *VersionsAndOperations*

$\vdash$    $\dfrac{inductive\_hyp \; : \; trans\_version\_reifctn \; (v_1, v_2, v_3)}{\langle\, \text{Proof of } trans\_version\_reifctn(op_1 \odot v_1, op_2 \odot v_2, op_3 \odot v_3). \; 5.6.6.5\,\rangle}$

$]$

This code is used in section 5.6.6.2.

**5.6.6.5.**   The transitivity proof can be further broken down in an obvious way.

$\langle\,$Proof of $trans\_version\_reifctn(op_1 \odot v_1, op_2 \odot v_2, op_3 \odot v_3). \; 5.6.6.5\,\rangle \equiv$

$[\, reif_{12}\_hyp \; : \; version\_reif(op_1 \odot v_1, op_2 \odot v_2, retr_{12})$

$\vdash[\, reif_{23}\_hyp \; : \; version\_reif(op_2 \odot v_2, op_3 \odot v_3, retr_{23})$

   $\vdash \langle\,$Proof of $version\_reif(op_1 \odot v_1, op_3 \odot v_3, retr_{13}) \; 5.6.6.6\,\rangle$

   $]$

     $\backslash\, imp. \; intro$

$]$

$\backslash\, imp. \; intro$

     $\therefore \; trans\_version\_reifctn(op_1 \odot v_1, op_2 \odot v_2, op_3 \odot v_3)$

This code is used in section 5.6.6.4.

**5.6.6.6.**   We now come to the core of the proof: the hard part (*newop*) consists of showing the operation reification condition for the operator pair $op_1$ and $op_3$. The idea is to use transitivity of operator reification, derived above (*op\_refine*). The easy part of the condition (*base*) follows from the inductive hypothesis.

$\langle\,$Proof of $version\_reif(op_1 \odot v_1, op_3 \odot v_3, retr_{13}) \; 5.6.6.6\,\rangle \equiv$

$\langle\! \! \langle\ base \quad := \; inductive\_hyp$

         $\backslash\, imp. \; elim \; (def\_version\_reif. \; cons. \; down \; (reif_{12}\_hyp).base)$

         $\backslash\, imp. \; elim \; (def\_version\_reif. \; cons. \; down \; (reif_{23}\_hyp).base)$

           $\therefore \; version\_reif(v_1, v_3, retr_{13})$

$, newop \; := \; op\_refine$

        $/ \; def\_version\_reif. \; cons. \; down \; (reif_{12}\_hyp). \; newop$

        $/ \; def\_version\_reif. \; cons. \; down \; (reif_{23}\_hyp). \; newop$

         $\therefore \; op_1 \sqsubseteq^{op}_{inv_1, retr_{13}} op_3$

$\langle\! \! \rangle$

$\backslash\, def\_version\_reif. \; cons. \; up$

     $\therefore \; version\_reif(op_1 \odot v_1, op_3 \odot v_3, retr_{13})$

This code is used in section 5.6.6.5.

### 5.6.7  Proof Assembly

Finally the above results can be put together to prove that version $v_1$ is a reification of version $v_3$ wrt. the retrieve function $retr_{13}$. Remember that, according to the definition of reification, there are three conditions to be satisfied: the version verification condition (which follows directly from the reification assumptions), the conditions for the composed retrieve function (which has been verified in Sect. 5.6.4), and the version reification condition (which follows from the derivations in Sect. 5.6.6) using the version reification conditions of the assumptions.

$$\langle \text{Proof assembly. } 5.6.7 \rangle \equiv$$

$$\langle version \quad := \langle concrete := reif_{12} \,.\, version \,.\, concrete$$

$$, abstract := reif_{23} \,.\, version \,.\, abstract$$

$$\rangle$$

$$\therefore \langle v_1 \checkmark, v_3 \checkmark \rangle$$

$$, retrieval \quad := retrieval_{13} \therefore valid\_retrieve(inv_1, inv_3, retr_{13})$$

$$, reification := trans\_version\_reifctn\_proof$$

$$\setminus imp.\ elim\ (reif_{12}.reification)$$

$$\setminus imp.\ elim\ (reif_{23}.reification)$$

$$\therefore version\_reif(v_1, v_3, retr_{13})$$

$$\rangle$$

This code is used in section 5.6.3.4.

## 5.7  Discussion

Currently the VDM methodology is applied predominantly in a semi-formal style. In this case study we have experimented with a complete formalization of VDM developments. We will try to evaluate this experiment at the end of the book, i.e., discuss drawbacks and benefits, also in comparison with the case study on algorithm calculation. At this place we would like to briefly discuss some phenomena that are more specific to the nature of VDM.

The formalization of VDM datastructures has been incomplete but typical. It was not intended to match the currently evolving VDM standardisation, however we hope that the reader is convinced enough that the given formalization could be extended to cover the VDM datastructures and notations to a very wide degree. A similar remark applies to the formalization of LPF and to the formalization of the HLA development. Furthermore, we did not tackle at all topics such as operation decomposition and modules in VDM.

Transitivity of VDM reification is neither deep nor trivial to prove. The proof has been elaborated in a mathematical style on paper, before becoming formalized in Deva. Proving properties like transitivity of reification helps to better understand formal methods. For example it is instructive to see, using the cross-reference tables at the end of the book, where and how precisely the condition

# 6 Case Study on Algorithm Calculation

This chapter presents a formalization of some selected parts of an algorithmic calculus for developing programs from specifications known as the Bird-Meertens Formalism or "Squiggol" [14], [78]. This calculus derives its power from recognizing some basic, yet powerful, laws that come with algebraic data types such as trees, lists or bags. In particular, these laws express the properties of a few higher order operators over these types. Together with an economic and concise, APL-like, notation, these laws allow the *calculation* of an efficient functional program from a (possibly) inefficient functional specification without resorting to inductive proofs.

The design of the Bird-Meertens formalism began with the development of a theory of lists centered around some select problem classes. However, the generality of this approach to algorithm calculation was quickly recognized and the calculus is now being actively extended by various groups: On the one hand, new theories for specific data types and specific problem classes are developed (cf., e.g., [58] and [15]) On the other hand, the calculus is redesigned in a relational setting (in contrast to the original functional one) in order to better cope with polymorphism and non-determinism (cf. [7] and [8]). A readily obtainable overview of recent developments is the paper by Malcolm [76]. He extends earlier work by Hagino [49] in categorical programming and shows how key transformation laws for (categorically defined) data types can be derived from their definition.

## 6.1 Overview

The formalization we want to present in this chapter covers some basic data structures, their properties, and the calculations of two algorithms. In fact, it is a formalization of what Bird describes in Secs. 1 and 3 of [14]. The basic theory for algorithm calculation consists, in the context of this setting, of a sorted predicate calculus with an extensional notion of equality and some elementary notions of algebra. See Chap. 4 for the definition of the context *CalculationalBasics*.

The formalization begins by introducing *join lists*, i.e., lists with constructors for the empty list, for the singleton list, and for the concatenation (or join) of two lists. By excluding the constructor for the empty list, one obtains non-empty join lists which will come up in Sect. 6.7. Some basic theory about join lists is then developed, in particular the "promotion theorem" is stated and two instances are derived. Some of the results are adapted to the case of non-empty join lists. After that, the notion of *segments* is introduced and a development law resembling the Horner-scheme for computing polynomials is presented. The given laws are then used to develop an efficient algorithm solving a segment problem: A linear time algorithm for computing the maximum segment sum of a list of integers. Finally, non-empty lists are used to introduce the data type of trees with finite but non-zero branching, and an important optimization of tree search algorithms is derived: it is shown how the $\alpha\beta$-*pruning* strategy for two player game-trees can be derived starting from the minimax evaluation scheme.

**context**  *CalculationalDevelopment* :=

[ **import**  *BasicTheories*

; **import**  *CalculationalBasics*

; ⟨ Join lists. 6.2 ⟩

; ⟨ Non-empty join lists. 6.3 ⟩

; ⟨ Some theory of join lists. 6.4 ⟩

; ⟨ Some theory of non-empty join lists. 6.5 ⟩

; ⟨ Segment problems. 6.6 ⟩

; ⟨ Tree evaluation problems. 6.7 ⟩

]

## 6.2　Join Lists

Join lists, just like the data types presented in Chap. 4, will be formalized by
first declaring the constructors, and then their defining axioms. In addition, we
define several higher-order operators on join lists.

⟨ Join lists. 6.2 ⟩ ≡

**context**  *JoinLists* :=

[ *list*  : [ *sort* ⊢ *sort* ]

; ⟨ Constructors of join lists. 6.2.1 ⟩

; ⟨ Axioms of join lists. 6.2.2 ⟩

; ⟨ Map and reduce operators for join lists. 6.2.3 ⟩

; ⟨ Left and right reduce operators for join lists. 6.2.4 ⟩

]

This code is used in section 6.1.

**6.2.1.**　Join lists are built from three constructors: The empty list $(1_{+\!\!\!+})$, the
singleton list constructor $(\langle\!\langle (\cdot) \rangle\!\rangle)$, and concatenation or "join" operation $(+\!\!\!+)$.
For example, the expression $\langle a \rangle +\!\!\!+ (\langle b \rangle +\!\!\!+ \langle c \rangle)$ is a list consisting of the three
elements $a$, $b$, and $c$, where the type declaration of $+\!\!\!+$ requires $a$, $b$, and $c$ to be
of the same sort. In the sequel, such right bracketed lists shall be abbreviated to
$\langle a, b, c \rangle$.

⟨ Constructors of join lists. 6.2.1 ⟩ ≡

$1_{+\!\!\!+}$　　　: [ $s$ ? *sort* ⊢ *list*$(s)$ ]

; $\langle\!\langle (\cdot) \rangle\!\rangle$　　: [ $s$ ? *sort* ⊢ [ $s$ ⊢ *list*$(s)$ ] ]

; $(\cdot) +\!\!\!+ (\cdot)$ : [ $s$ ? *sort* ⊢ [ *list* $(s)$; *list*$(s)$ ⊢ *list*$(s)$ ] ]

This code is used in section 6.2.

**6.2.2.** The singleton list constructor is an injection and the join operation is associative with the empty list as identity, i.e., join lists have a monoid structure. In fact, they are the *free* monoid structure over the base sort and this property is characterized by an induction axiom.

⟨ Axioms of join lists. 6.2.2 ⟩ ≡

   $psingleton$      : $injective\ (\langle\Box\rangle)$

 ; $pjoin$           : $monoid\ (+\!\!+, 1_{+\!\!+})$

 ; $join\_induction$ :

  $[\, s\ ?\ sort\ ;\ P\ ?\ [\, list\ (s) \vdash prop\,]$

$$\vdash\ \dfrac{P\,(1_{+\!\!+});\, [\, x : s\ \vdash P(\langle x\rangle)\,];\, [xs, ys\ ?\ list\ (s)\ \vdash \dfrac{P\,(xs);\, P(ys)}{P\,(xs +\!\!+ ys)}\,]}{[\,xs\ ?\ list\ (s) \vdash P(xs)\,]}$$

  $]$

This code is used in section 6.2.

**6.2.3.** In the context of algorithm calculation, there are two very important operators on join lists: *map* and *reduce*. The map operator ($*$) applies a function to each element of a list and returns the list of results, i.e.,

$$f * \langle a_1, \ldots, a_n\rangle \quad = \quad \langle f(a_1), \ldots, f(a_n)\rangle.$$

Mapping a function over the empty list returns the empty list which makes the "section" $f*$ a homomorphism on the monoid of join lists. The reduce operator '$/$', in effect, "feeds" the elements of a list to a monoid operator '$\oplus$', i.e.,

$$\oplus/\langle a_1, \ldots, a_n\rangle \quad = \quad a_1 \oplus \cdots \oplus a_n.$$

Reducing the empty list by a monoid operator returns the unit of the operator which again makes the section $\oplus/$ a homomorphism on the monoid of join lists. These two operators are now formally specified in Deva by the following declarations.

⟨ Map and reduce operators for join lists. 6.2.3 ⟩ ≡

  $(\cdot) * (\cdot)$     : $[s, t\ ?\ sort\ \vdash [[s \vdash t];\, list(s) \vdash list(t)]\,]$

 ; $map\_def$    : $[s, t\ ?\ sort\ ;\ f\ ?\ [s \vdash t]$

               $\vdash\!\{\ empty$      := $f * 1_{+\!\!+} = 1_{+\!\!+}$

                 $, singleton$ := $[\, x\ ?\ s\ \vdash f * \langle x\rangle = \langle f(x)\rangle\,]$

                 $, join$        := $[xs, ys\ ?\ list\ (s) \vdash f * (xs +\!\!+ ys)$        $]$

                                       $= (f * xs) +\!\!+ (f * ys)$

              $\flat$

         $]$

$; (\cdot) \,/\, (\cdot) \qquad : [\, s \,?\, sort \,\vdash\, [[s; s \vdash s]; \mathit{list}(s) \vdash s]]$
$; \mathit{reduce\_def} \, :$
$[\, s \,?\, sort \,;\, (\cdot) \oplus (\cdot) \,?\, [s; s \vdash s]; 1_{\oplus} \,?\, s$

$$\mathit{monoid}\ (\oplus, 1_{\oplus})$$

$$
\begin{aligned}
&\langle\!\!\langle\ empty \quad := (\oplus)\,/\,1_{+\!+} = 1_{\oplus} \\
&\vdash\!\!\dashv\, ,\ singleton := [\, x \,?\, s \,\vdash\, (\oplus)\,/\,\langle x \rangle = x\,] \\
&\quad\ ,\ join \qquad := [\, xs, ys \,?\, \mathit{list}\,(s) \vdash (\oplus)\,/\,(xs +\!\!+ ys) = ((\oplus)\,/\,xs) \oplus ((\oplus)\,/\,ys)] \\
&\,\rangle\!\!\rangle
\end{aligned}
$$

$]$

This code is used in section 6.2.

**6.2.4.** There are two more related reduce operators for lists: left-reduce '$\twoheadrightarrow$' and right reduce '$\twoheadleftarrow$' which feed the elements of a list to any binary operator in a fixed order starting with some seed, i.e.,

$$
\begin{aligned}
\oplus \twoheadrightarrow_a \langle a_1, \ldots, a_n \rangle &= ((a \oplus a_1) \oplus a_2) \cdots \oplus a_n, \\
\oplus \twoheadleftarrow_a \langle a_1, \ldots, a_n \rangle &= a_1 \oplus \cdots (a_{n-1} \oplus (a_n \oplus a)).
\end{aligned}
$$

For an illustration of such a *directed reduction*, consider the following equation which reminds one of Horner's rule for computing polynomials.

$$(a_1 \times a_2 \times \cdots \times a_n) + (a_2 \times a_3 \times \cdots \times a_n) + \cdots + a_n + 1$$
$$= \quad \{\text{ multiplication distributes over addition }\}$$
$$(((a_1 + 1) \times a_2 + 1) \times \cdots) \times a_n + 1$$

The latter expression can be expressed in terms of a left-reduction:

$$(((a_1 + 1) \times a_2 + 1) \times \cdots) \times a_n + 1$$
$$= \quad \{\text{ define } \circledast \text{ by } a \circledast b \triangleq a \times b + 1\ \}$$
$$(((1 \circledast a_1) \circledast a_2) \circledast \cdots) \circledast a_n$$
$$= \quad \{\text{ definition of } \twoheadrightarrow \ \}$$
$$\circledast \twoheadrightarrow_1 \langle a_1, \ldots, a_n \rangle$$

The defining equations for left- and right-reduction complete the formalization of join lists.

$\langle$ Left and right reduce operators for join lists. 6.2.4 $\rangle \equiv$

$\quad (\cdot) \twoheadrightarrow_{(\cdot)} (\cdot) \quad : [s, t \,?\, sort \,\vdash\, [[t; s \vdash t]; t; \mathit{list}(s) \vdash t]]$
$; (\cdot) \twoheadleftarrow_{(\cdot)} (\cdot) \quad : [s, t \,?\, sort \,\vdash\, [[s; t \vdash t]; t; \mathit{list}(s) \vdash t]]$
$; \mathit{lreduce\_def} \, :$

$[s, t ? sort ; (\cdot) \oplus (\cdot) ? [t; s \vdash t]; x ? t$
$\vdash\!\{ base := (\oplus) \not\!\not\to_x 1_{+\!+} = x$
$\quad , rec := [ y ? s ; ys ? list (s) \vdash (\oplus) \not\!\not\to_x (ys +\!+ \langle y \rangle) = ((\oplus) \not\!\not\to_x ys) \oplus y ]$
$\quad \flat$
$]$
$; rreduce\_def :$
$[s, t ? sort ; (\cdot) \oplus (\cdot) ? [s; t \vdash t]; x ? t$
$\vdash\!\{ base := (\oplus) \not\!\not\leftarrow_x 1_{+\!+} = x$
$\quad , rec := [ y ? s ; ys ? list (s) \vdash (\oplus) \not\!\not\leftarrow_x (\langle y \rangle +\!+ ys) = y \oplus ((\oplus) \not\!\not\leftarrow_x ys) ]$
$\quad \flat$
$]$

This code is used in section 6.2.

## 6.3  Non-empty Join Lists

Non-empty join lists are obtained from join lists by removing the constructor of
the empty list and suitably adapting all further declarations. This adaptation is
straightforward and the result is listed below. The reader may compare with the
corresponding declarations for (ordinary) join lists.

$\langle$ Non-empty join lists. 6.3 $\rangle \equiv$
**context** $NonEmptyJoinLists :=$
$[\![$ ne\_list : [sort \vdash sort]
$; \langle$ Constructors of non-empty join lists. 6.3.1 $\rangle$
$; \langle$ Axioms of non-empty join lists. 6.3.2 $\rangle$
$; \langle$ Map operator for non-empty join lists. 6.3.3 $\rangle$
$; \langle$ Reduce operator for non-empty join lists. 6.3.4 $\rangle$
$; \langle$ Left and right reduce operators for non-empty join lists. 6.3.5 $\rangle$
$]\!]$

This code is used in section 6.1.

**6.3.1.**    $\langle$ Constructors of non-empty join lists. 6.3.1 $\rangle \equiv$
$\langle (\cdot) \rangle$      $: [ s ? sort \vdash [s \vdash ne\_list(s)]]$
$; (\cdot) +\!+ (\cdot) : [ s ? sort \vdash [ ne\_list (s); ne\_list(s) \vdash ne\_list(s)]]$

This code is used in section 6.3.

**6.3.2.**    ⟨Axioms of non-empty join lists. 6.3.2⟩ ≡

$psingleton$          : $injective\ (\langle\Box\rangle)$

; $pjoin$               : $associative\ (+\!\!+)$

; $join\_induction$ :

  $[\,s\ ?\ sort\,;\ P\ ?\,[\,ne\_list\,(s)\vdash prop\,]$

$$\vdash \left[\,[\,x:s\vdash P(\langle x\rangle)\,]\,;\,[\,xs,ys\ ?\ ne\_list\,(s)\vdash \frac{P\,(xs);\,P(ys)}{P\,(xs+\!\!+\,ys)}\,]\right.$$

$$[\,xs\ ?\ ne\_list\,(s)\vdash P(xs)\,]$$

$]$

This code is used in section 6.3.

**6.3.3.**    ⟨Map operator for non-empty join lists. 6.3.3⟩ ≡

$(\cdot) * (\cdot)$    : $[\,s,t\ ?\ sort\vdash [[\,s\vdash t\,];\ ne\_list(s)\vdash ne\_list(t)\,]\,]$

; $map\_def$ :

$[\,s,t\ ?\ sort\,;\ f\ ?\,[\,s\vdash t\,]$

$\vdash\langle\ singleton := [\,x\ ?\ s\vdash f*\langle x\rangle = \langle f(x)\rangle\,]$

$,\ join$        $:= [\,xs,ys\ ?\ ne\_list\,(s)\vdash f*(xs+\!\!+ys) = (f*xs)+\!\!+(f*ys)\,]$

$\rangle$

$]$

This code is used in section 6.3.

**6.3.4.**    ⟨Reduce operator for non-empty join lists. 6.3.4⟩ ≡

$(\cdot) / (\cdot)$      : $[\,s\ ?\ sort\vdash [[\,s;s\vdash s\,];\ ne\_list(s)\vdash s\,]\,]$

; $reduce\_def$ :

$[\,s\ ?\ sort\,;\ (\cdot)\oplus(\cdot)\ ?\,[\,s;s\vdash s\,];\ 1_\oplus\ ?\ s$

$$associative\ (\oplus)$$

$\vdash\langle\ singleton := [\,x\ ?\ s\vdash(\oplus)/\langle x\rangle = x\,]$

$,\ join$        $:= [\,xs,ys\ ?\ ne\_list\,(s)\vdash(\oplus)/(xs+\!\!+ys)$

$$= ((\oplus)/xs)\oplus((\oplus)/ys)$$

$\rangle$

$]$

This code is used in section 6.3.

**6.3.5.**    ⟨Left and right reduce operators for non-empty join lists. 6.3.5⟩ ≡

$(\cdot)\not\to_{(\cdot)}(\cdot)$    : $[\,s,t\ ?\ sort\vdash [[\,t;s\vdash t\,];\ t;\ ne\_list(s)\vdash t\,]\,]$

; $(\cdot)\not\!\leftarrow_{(\cdot)}(\cdot)$    : $[\,s,t\ ?\ sort\vdash [[\,s;t\vdash t\,];\ t;\ ne\_list(s)\vdash t\,]\,]$

; *lreduce_def* :

$[s, t\ ?\ sort\ ; (\cdot) \oplus (\cdot)\ ?\ [t; s \vdash t]\ ;\ x\ ?\ t\ ;\ y\ ?\ s$
$\vdash \langle\ base\ := (\oplus) \not\nearrow_x \langle y \rangle = x \oplus y$
$\quad,\ rec\quad := [\ ys\ ?\ ne\_list\ (s) \vdash (\oplus) \not\nearrow_x (ys + \langle y \rangle) = ((\oplus) \not\nearrow_x ys) \oplus y\,]$
$\quad \rangle$
]

; *rreduce_def* :

$[s, t\ ?\ sort\ ; (\cdot) \oplus (\cdot)\ ?\ [s; t \vdash t]\ ;\ x\ ?\ t\ ;\ y\ ?\ s$
$\vdash \langle\ base\ := (\oplus) \not\swarrow_x \langle y \rangle = y \oplus x$
$\quad,\ rec\quad := [\ ys\ ?\ ne\_list\ (s) \vdash (\oplus) \not\swarrow_x (\langle y \rangle + ys) = y \oplus ((\oplus) \not\swarrow_x ys)\,]$
$\quad \rangle$
]

This code is used in section 6.3.

We have chosen to introduce non empty join lists in separation from join lists. Alternatively, it would have been possible to first introduce non-empty join lists, and then obtain (arbitrary) join lists by adding the empty list and extending the defining laws of the operators. While technically being perfectly feasible, this approach leads to a somewhat implicit formalization of join lists, emphasizing very much the differences to non-empty join lists. Since join lists are one of the "classical" datastructures of calculational algorithm calculation, we have chosen to follow the approach to explicitly formalize them in the first place.

Note that in Deva it is not possible to obtain non-empty join lists by "hiding" the empty list in join lists. The problem is that many laws of join lists involve the empty list, and it is not clear how these laws should be adapted in general. Note further that we used identical names for the operators and laws of both kinds of lists. Thus name-clashes would arise when importing in parallel the contexts formalizing the two types of lists. These name clashes would have to be resolved by renaming, as done for example in 4.4.1. However, in this chapter the two types of lists will not be used at the same time.

## 6.4   Some Theory of Join Lists

⟨ Some theory of join lists. 6.4 ⟩ ≡
**context**  *TheoryOfJoinLists* :=
[ **import**  *JoinLists*
; ⟨ Map distribution. 6.4.1 ⟩
; ⟨ Catamorphisms. 6.4.2 ⟩
; ⟨ Promotion theorems. 6.4.3 ⟩
]

This code is used in section 6.1.

**6.4.1.** One of the properties of the map operator which can be derived from the laws given in the previous chapter is that it distributes over composition, i.e.,

$$(f*) \circ (g*) \;=\; (f \circ g) * .$$

for functions $f$ and $g$ of proper type. An inductive proof of this property is left as an easy exercise for the reader. Since its formalization would somewhat distract from the line of this presentation, it is omitted. For examples of formalized proofs by join list induction in Deva see [104].

⟨ Map distribution. 6.4.1 ⟩ ≡

*map_distribution* :

$[s, t, u \,?\;\; sort \,; f \,?\; [t \vdash u]; g \,?\; [s \vdash t] \vdash (f*) \circ (g*) = (f \circ g) * \,]$

This code is used in section 6.4.

**6.4.2.** A homomorphism whose domain is the monoid of join lists is called a *(join-list) catamorphism*. A central theorem of the theory of lists states that any catamorphism can be uniquely factorized into a reduce and a map, i.e., for any catamorphism $h$ there exist a monoid operator $\oplus$ with identity $1_\oplus$ and a function $f$ such that

$$h \;=\; (\oplus/) \circ (f*),$$

i.e.,

$$
\begin{aligned}
h(1_{\!+\!\!+}) &= 1_\oplus, \\
h(\langle a \rangle) &= f(a), \\
h(x +\!\!+ y) &= h(x) \oplus h(y).
\end{aligned}
$$

Because this factorization is unique, we can introduce a special notation to denote catamorphisms: $([\oplus, f]) \triangleq (\oplus/) \circ (f*)$. A further consequence is that any catamorphism can be written as a left- or a right reduction

$$
([\oplus, f]) \;=\;
\begin{cases}
\circledast \!\not\to\!_{1_\oplus}, & \text{where } a \circledast b \triangleq a \oplus f(b) \\
\circledast \!\not\leftarrow\!_{1_\oplus}, & \text{where } a \circledast b \triangleq f(a) \oplus b
\end{cases}
$$

Proofs of these *specialization* properties of catamorphisms are again left to the reader as an exercise, we just state the corresponding declarations:

⟨ Catamorphisms. 6.4.2 ⟩ ≡

$([(\cdot), (\cdot)]) \quad := [s, t \,?\; sort \vdash [\oplus : [t; t \vdash t]; f : [s \vdash t] \vdash (\oplus /) \circ (f*)]]$

; *specialization* :

$[s, t ?\ sort\ ; (\cdot) \oplus (\cdot)\ ?\ [t; t \vdash t]; 1_\oplus\ ?\ t\ ; f\ ?\ [s \vdash t]$

$$monoid\ (\oplus, 1_\oplus)$$

$\vdash$ $\langle left\ \ := [\ \circledast := [\ a : t\ ; b :\ s\ \vdash a \oplus f(b)\ ] \vdash ([\oplus, f]) = \circledast \not{+}_{1_\oplus}\ ]$
$, right := [\ \circledast := [\ a : s\ ; b :\ t\ \vdash f(a) \oplus b\ ] \vdash ([\oplus, f]) = \circledast \not{+}_{1_\oplus}\ ]$
$\rangle$

$]$

This code is used in section 6.4.

**6.4.3.**   A central notion of algorithm calculation is promotability: A function $f$ is $\oplus \to \otimes$-*promotable*, where both binary operators are assumed to be monoid operators, iff

$$\begin{aligned} f(1_\oplus) &= 1_\otimes \\ f(a \oplus b) &= f(a) \otimes f(b) \end{aligned}$$

This says nothing but that $f$ is a monoid homomorphism, the purpose of the notation is to emphasize how $f$ behaves with respect to the monoid operators. Clearly, any catamorphism $([\oplus, f])$ is $(\mathbin{+\mkern-8mu+} \to \oplus)$-promotable. One is interested in the promotability properties of a function because they imply powerful transformation laws. This is stated by the *promotion theorem*: A function $f$ is $\oplus \to \otimes$-promotable, iff

$$f \circ (\oplus/) = (\otimes/) \circ (f*)$$

The equivalence of these two characterizations of promotability is formalized in Deva as follows.

$\langle$ Promotion theorems. 6.4.3 $\rangle \equiv$

$promotion : [s, t ?\ sort\ ; (\cdot) \oplus (\cdot)\ ?\ [s; s \vdash s]; 1_\oplus\ ?\ s\ ; (\cdot) \otimes (\cdot)\ ?\ [t; t \vdash t]; 1_\otimes\ ?\ t$

$\langle\ monoid\ (\oplus, 1_\oplus), monoid\ (\otimes, 1_\otimes)\rangle; f : [s \vdash t]$

$\vdash \langle\!\langle f\ (1_\oplus) = 1_\otimes, [a, b\ ?\ s\ \vdash f(a \oplus b) = f(a) \otimes f(b)]\rangle\!\rangle$

$$f \circ ((\oplus)\ /) = ((\otimes)\ /) \circ (f*)$$

$]$

See also sections 6.4.4 and 6.4.5.

This code is used in section 6.4.

**6.4.4.**   We can use the promotion theorem to prove two theorems of the theory of lists: *map promotion*

$$(f*) \circ (\mathbin{+\mkern-8mu+}/) = (\mathbin{+\mkern-8mu+}/) \circ ((f*)*),$$

i.e., $f*$ is $+\!\!\!+ \to +\!\!\!+$-promotable, and *reduce promotion*

$$(\oplus/) \circ (+\!\!\!+/) \quad = \quad (\oplus/) \circ ((\oplus/)*),$$

i.e., $\oplus/$ is $+\!\!\!+ \to \oplus$-promotable, under the assumption that $\oplus$ is a monoid operator. Intuitively, these two laws correspond to the equations

$$(f*)(x_1 +\!\!\!+ \cdots +\!\!\!+ x_n) \quad = \quad (f * x_1) +\!\!\!+ \cdots +\!\!\!+ (f * x_n), \quad \text{and}$$
$$(\oplus/)(x_1 +\!\!\!+ \cdots +\!\!\!+ x_n) \quad = \quad (\oplus/x_1) \oplus \cdots \oplus (\oplus/x_n).$$

Both theorems are obtained as simple instantiations of the promotion theorem.

⟨ Promotion theorems. 6.4.3 ⟩$+ \equiv$
; *map_promotion*   :=
$[s, t ? \; sort \,; f \; ? \, [s \vdash t]$
$\vdash promotion$
$\quad (\langle\!| \; pjoin \,, pjoin \;|\!\rangle, (f*)). \; down \; (\langle\!| \; map\_def . \, empty \,, map\_def . join \;|\!\rangle)$
$]$
$\quad\quad \therefore [s, t ? \; sort \,; f \; ? \, [s \vdash t] \vdash (f*) \circ ((+\!\!\!+) /) = ((+\!\!\!+) /) \circ ((f*)*) \,]$

; *reduce_promotion* :=
$[s \; ? \; sort \,; \oplus \; ? \, [s; s \vdash s] ; 1_\oplus \; ? \; s$
$\vdash [ \; plus\_props \; : \; monoid \, (\oplus, 1_\oplus)$
$\quad \vdash promotion \; (\langle\!| \; pjoin \,, plus\_props \;|\!\rangle, ((\oplus) /)). \; down$
$\quad\quad (\langle\!| \; reduce\_def \, (plus\_props). \, empty \,, reduce\_def \, (plus\_props). join \;|\!\rangle)$
$\quad ]$
$]$

$$\therefore [\, s ? \; sort \,; \oplus ? \, [s; s \vdash s]; 1_\oplus \; ? \; s \vdash \left| \frac{monoid \, (\oplus, 1_\oplus)}{((\oplus) /) \circ ((+\!\!\!+) /) = ((\oplus) /) \circ (((\oplus) /)*)} \right]$$

**6.4.5.** We combine these two laws in form of a simple promotion tactic which, when used, tries to unfold with one of them. It will be used in Sect. 6.6. Note that this tactic is conditional, i.e., it depends on an additional argument, namely $props_\oplus$ which is required by the reduce promotion law.

⟨ Promotion theorems. 6.4.3 ⟩$+ \equiv$
; *promotion_tac* :=
$[s \; ? \; sort \,; \oplus \; ? \, [s; s \vdash s] ; 1_\oplus \; ? \; s$
$\vdash [ \; plus\_props \; : \; monoid \, (\oplus, 1_\oplus)$
$\quad \vdash \langle\!| \; unfold \, (map\_promotion), \; unfold \, (reduce\_promotion(plus\_props)) \;|\!\rangle$
$\quad ]$
$]$

## 6.5 Some Theory of Non-Empty Join Lists

A theory of non-empty join lists, sufficient for the purposes of this case study, is obtained by selection and adaptation of the material of the previous section. Essentially everything involving empty lists and identity elements of binary operators is removed. The specialization rule is stated in a stronger form, i.e., with a weaker condition: the initialization element $e$ of a left-reduction $\circledast \not\!/_e$ has to be an identity on the range of $f$ only. This stronger form is useful for the example in Sect. 6.7.

$\langle$ Some theory of non-empty join lists. $6.5 \rangle \equiv$

**context** *TheoryOfNonEmptyJoinLists* :=

$[\![$ **import** *NonEmptyJoinLists*

; $\langle$ Map distribution (for non-empty join lists). $6.5.1 \rangle$

; $\langle$ Catamorphisms (for non-empty join lists). $6.5.2 \rangle$

; $\langle$ Promotion theorem (for non-empty join lists). $6.5.3 \rangle$

$]\!]$

This code is used in section 6.1.

**6.5.1.** $\langle$ Map distribution (for non-empty join lists). $6.5.1 \rangle \equiv$

*map_distribution* :

$[s, t, u ?\ sort\ ; f\ ?\ [t \vdash u]; g\ ?\ [s \vdash t] \vdash (f*) \circ (g*) = (f \circ g) * ]$

This code is used in section 6.5.

**6.5.2.** $\langle$ Catamorphisms (for non-empty join lists). $6.5.2 \rangle \equiv$

$(\![ (\cdot), (\cdot) ]\!)$ $\quad := [s, t ?\ sort \vdash [\ \oplus :[t; t \vdash t]; f : [s \vdash t] \vdash (\oplus\,/) \circ (f*)]]$

; *specialization* :

$[s, t ?\ sort\ ;\ (\cdot) \oplus (\cdot)\ ?\ [t; t \vdash t];\ e\ ?\ t;\ f\ ?\ [s \vdash t]$

$\quad \big|\ associative\ (\oplus); [\ b\ ?\ s \vdash ((e \oplus f(b)) =)(f(b))]$

$\vdash \Big| \langle left\ \ := [\ \circledast := [\ a : t\ ; b : s \vdash a \oplus f(b)] \vdash (\![ \oplus, f ]\!) = \circledast \not\!/_e\ ]$

$\quad , right := [\ \circledast := [\ a : s\ ; b : t \vdash f(a) \oplus b] \vdash (\![ \oplus, f ]\!) = \circledast \not\!\backslash_e\ ]$

$\quad \rangle$

$]$

This code is used in section 6.5.

**6.5.3.** $\langle$ Promotion theorem (for non-empty join lists). $6.5.3 \rangle \equiv$

*promotion* :

$[s, t \ ? \ sort \ ; \ (\cdot) \oplus (\cdot) \ ? \ [s; s \vdash s]; \ (\cdot) \otimes (\cdot) \ ? \ [t; t \vdash t]$
$\| \ \| \langle \ associative \ (\oplus), \ associative \ (\otimes) \rangle; f : [s \vdash t]$
$\vdash \| [a, b \ ? \ s \vdash f(a \oplus b) = f(a) \otimes f(b)]$
$\| \qquad f \circ ((\oplus) \ /) = ((\otimes) \ /) \circ (f*)$
$]$

This code is used in section 6.5.

## 6.6   Segment Problems

The first development presented in this chapter is taken from a class of problems involving segments (i.e., contiguous sublists). Before presenting the problem and its solution, some theory about segments is necessary.

$\langle$ Segment problems. 6.6 $\rangle \equiv$

**context** *SegmentProblems* :=
$[$ **import**  *TheoryOfJoinLists*
; $\langle$ Initial and final segments. 6.6.1 $\rangle$
; $\langle$ Horner's rule. 6.6.2 $\rangle$
; $\langle$ Left and right accumulations. 6.6.3 $\rangle$
; $\langle$ The "maximum segment sum" problem. 6.6.4 $\rangle$
$]$

This code is used in section 6.1.

**6.6.1.**   *Initial* and *final segments* of a list are recursively defined by

$$
\begin{aligned}
inits(1_{+\!\!+}) &= \langle 1_{+\!\!+} \rangle, \\
inits(\langle a \rangle +\!\!+ x) &= \langle 1_{+\!\!+} \rangle +\!\!+ (((\langle a \rangle +\!\!+) * inits(x)) \\
tails(1_{+\!\!+}) &= \langle 1_{+\!\!+} \rangle, \\
tails(x +\!\!+ \langle a \rangle) &= ((+\!\!+ \langle a \rangle) * tails(x)) +\!\!+ \langle 1_{+\!\!+} \rangle
\end{aligned}
$$

i.e.,

$$
\begin{aligned}
inits(\langle a_1, \ldots, a_n \rangle) &= \langle 1_{+\!\!+}, \langle a_1 \rangle, \langle a_1, a_2 \rangle, \ldots, \langle a_1, \ldots, a_n \rangle \rangle, \\
tails(\langle a_1, \ldots, a_n \rangle) &= \langle \langle a_1, \ldots, a_n \rangle, \langle a_2, \ldots, a_n \rangle, \ldots, \langle a_n \rangle, 1_{+\!\!+} \rangle
\end{aligned}
$$

The *segments* of a list are obtained by first taking the list of initial segments, then forming the tails of each initial segment, and finally joining the resulting three-level list into a two-level list:

$$
segs \quad = \quad (+\!\!+/) \circ (tails*) \circ inits.
$$

Note that due to the fact that *tails* generates an empty list for any input, *segs* will, in general, generate multiple occurrences of the empty list.

$\langle$ Initial and final segments. 6.6.1 $\rangle \equiv$

$inits, tails: \quad [\, s \,?\, sort \,;\, list(s) \vdash list(list(s)) \,]$

$;\, def\_inits: \quad \langle\!| \; empty := inits\,(1_{+\!\!+}) = \langle 1_{+\!\!+} \rangle$

$\qquad\qquad\qquad , \; cons \quad := [\, s \,?\, sort \,;\, x \,?\, s \,;\, xs \,?\, list\,(s)$

$\qquad\qquad\qquad\qquad\qquad \vdash inits\,(\langle x \rangle +\!\!+ xs) = \langle 1_{+\!\!+} \rangle +\!\!+ ((\langle x \rangle +\!\!+) * inits(xs))$

$\qquad\qquad\qquad\qquad ]$

$\qquad\qquad\qquad \natural$

$;\, def\_tails: \quad \langle\!| \; empty := inits\,(1_{+\!\!+}) = \langle 1_{+\!\!+} \rangle$

$\qquad\qquad\qquad , \; cons \quad := [\, s \,?\, sort \,;\, x \,?\, s \,;\, xs \,?\, list\,(s)$

$\qquad\qquad\qquad\qquad\qquad \vdash inits\,(xs +\!\!+ \langle x \rangle) = ((+\!\!+ \langle x \rangle) * inits(xs)) +\!\!+ \langle 1_{+\!\!+} \rangle$

$\qquad\qquad\qquad\qquad ]$

$\qquad\qquad\qquad \natural$

$;\, segs \qquad := [\, s \,?\, sort \vdash ((+\!\!+)\, /)\, \circ\, (tails*)\, \circ\, inits \,]$

This code is used in section 6.6.

**6.6.2.** Using *tails*, we are now able to give a concise formulation of Horner's rule

$$(a_1 \times a_2 \times \cdots \times a_n) + (a_2 \times a_3 \times \cdots \times a_n) + \cdots + a_n + 1$$

$$=$$

$$(((a_1 + 1) \times a_2 + 1) \times \cdots) \times a_n + 1$$

Recall (p. 184) that the latter expression can be denoted as

$$\circledast \not\!/_1 \langle a_1, \dots, a_n \rangle,$$

where $a \circledast b \triangleq (a \times b) + 1$. For the first expression, we calculate:

$$(a_1 \times \cdots \times a_n) + (a_2 \times \cdots \times a_n) + \cdots + a_n + 1$$

$=$ { definition of $+/$ }

$\qquad +/\langle a_1 \times a_2 \times \cdots \times a_n, a_2 \times \cdots \times a_n, \dots, a_n, 1 \rangle$

$=$ { definition of $\times/$ }

$\qquad +/\langle \times/\langle a_1, \dots, a_n \rangle, \times/\langle a_2, \dots, a_n \rangle, \dots, \times/\langle a_n \rangle, \times/1_{+\!\!+} \rangle$

$=$ { definition of $(\times/)*$ }

$\qquad ((+/) \circ (\times/)*)\langle\langle a_1, \dots, a_n \rangle, \langle a_2, \dots, a_n \rangle, \dots, \langle a_n \rangle, 1_{+\!\!+} \rangle$

$=$ { definition of *tails* }

$\qquad ((+/) \circ (\times/)* \circ tails)\langle a_1, \dots, a_n \rangle$

By abstracting from the concrete operations $+$ and $\times$, Horner's rule can now be formulated completely inside the calculus: Assume that $\oplus$ and $\otimes$ are monoid operators and that the following distributivity property holds: $a \otimes (b \oplus c) = (a \otimes b) \oplus (a \otimes c)$. Then, Horner's rule states

$$(\oplus/) \circ (\otimes/)* \circ tails \quad = \quad \circledast \mathbin{/\!\!\!/}_{1_\otimes} ,$$

where $a \circledast b \triangleq (a \otimes b) \oplus 1_\otimes$. The Deva formalization is a direct transcription of the above formulation.

⟨ Horner's rule. 6.6.2 ⟩ ≡

*horner_rule* :

$[\, s \; ? \; sort \; ; \oplus, \otimes \; ? \; [s; s \vdash s]; 1_\oplus, 1_\otimes \; ? \; s \; ; \; \circledast \; := [a, b : s \vdash \oplus(\otimes(a, b), 1_\otimes)]$

$\vdash \Big|\quad \dfrac{distrib\_monoids\ (\oplus, 1_\oplus, \otimes, 1_\otimes)}{(\oplus \, /) \circ ((\otimes \, /)*) \circ tails = \circledast \mathbin{/\!\!\!/}_{1_\otimes}}$

$]$

This code is used in section 6.6.

**6.6.3.** Finally, two *accumulation* operators, whose purpose is to construct lists of successive intermediate results of directed reductions, can be formulated in terms of the directed reduction operators and initial segments. The operators of left-accumulation $\oplus \mathbin{/\!\!\!/}_a$ and right-accumulation $\oplus \mathbin{/\!\!/}_a$ are characterized by the equations:

$$\oplus \mathbin{/\!\!\!/}_a \quad = \quad (\oplus \mathbin{/\!\!/}_a)* \circ inits, \quad \text{and}$$
$$\oplus \mathbin{/\!\!/}_a \quad = \quad (\oplus \mathbin{/\!\!\!/}_a)* \circ tails,$$

or, more descriptively, by

$$\oplus \mathbin{/\!\!\!/}_a \langle a_1, \ldots, a_n \rangle \quad = \quad \langle a, a \oplus a_1, \ldots, ((a \oplus a_1) \oplus a_2) \oplus \cdots \oplus a_n \rangle, \quad \text{and}$$
$$\oplus \mathbin{/\!\!/}_a \langle a_1, \ldots, a_n \rangle \quad = \quad \langle a_1 \oplus \cdots \oplus (a_{n-1} \oplus (a_n \oplus a)), \ldots, a_n \oplus a, a \rangle$$

Note that, assuming $\oplus$ can be computed in constant time, both expressions on the right-hand side can be computed in linear time depending on the length of the input list. The formulation in Deva is again a straightforward transcription:

⟨ Left and right accumulations. 6.6.3 ⟩ ≡

$(\cdot) \mathbin{/\!\!\!/}_{(\cdot)} (\cdot) := [s, t \; ? \; sort \vdash [\oplus :[t; s \vdash t]; a : t \vdash ((\oplus \mathbin{/\!\!/}_a)*) \circ inits\,]]$

$; (\cdot) \mathbin{/\!\!/}_{(\cdot)} (\cdot) := [s, t \; ? \; sort \vdash [\oplus :[s; t \vdash t]; a : t \vdash ((\oplus \mathbin{/\!\!\!/}_a)*) \circ tails\,]]$

This code is used in section 6.6.

**6.6.4.** The theory presented in the last section will now be applied to develop an efficient algorithm for the maximum segment sum problem. The problem is, for a given list of integers, to compute the maximum of the sums of its segments. For example, the maximum segment sum (or *mss*) of the list $\langle -1, 3, -1, 2, -4, 3 \rangle$ is 4, the sum of the segment $\langle 3, -1, 2 \rangle$.

A first systematic approach to the solution is to proceed in three steps:

Step A: Compute all segments,
Step B: compute all their sums, and finally
Step C: compute the maximum of all these sums.

In the algorithm calculation style this specification-like algorithm is written as follows ('$\uparrow$' denotes the maximum operator):

$$mss \quad \triangleq \quad \underbrace{(\uparrow /)}_{C} \circ \underbrace{(+/)*}_{B} \circ \underbrace{segs}_{A}$$

Note, that $+/$ computes the sum of a list, and thus $(+/)*$ computes the list of sums of a list of lists. Intuitively, this algorithm is cubic, since it maps a linear function $(+/)$ over a list consisting of a quadratic number of lists. This first, rather natural but computationally expensive solution is therefore inappropriate to be considered as an implementation. A second, more constrained attempt is to proceed as follows:

Step A': Go through the elements of the list from left-to-right,
Step B': for every element, compute the maximum segment sum of all segments ending in that element, and finally
Step C': compute the maximum of all these sums.

This algorithm is written as follows:

$$mss \quad \triangleq \quad \underbrace{(\uparrow /)}_{C'} \circ \underbrace{((\uparrow /) \circ (+/)* \circ tails)*}_{B'} \circ \underbrace{inits}_{A'}$$

The nice property of this expression is that Horner's rule can be applied to B'.

The following proof in the calculus first shows the equivalence of the two solutions using tactics for promotion and distribution, and then transforms the second solution via Horner's rule to yield a linear algorithm. The laws necessary for the development are associativity (of both operators '$\uparrow$' and '+'), existence of identities ($\bot$ for $\uparrow$, 0 for +), and the distributivity property $(a \uparrow b) + c = (a + c) \uparrow (b + c)$. Consequently, we have a distributive monoid pair.

$\langle$ The "maximum segment sum" problem. 6.6.4 $\rangle \equiv$

| | | |
|---|---|---|
| $s$ | : | *sort* |
| $; (\cdot) \uparrow (\cdot), (\cdot) + (\cdot)$ | : | $[s; s \vdash s]$ |
| $; max\_plus\_props$ | : | $distrib\_monoids\ ((\uparrow), \bot, (+), 0)$ |
| $; max\_props$ | := | $max\_plus\_props \ . \ mon\_plus$ |

; *mss*              $:= ((\uparrow) /) \circ (((+) /)*) \circ segs \therefore [\, list\,(s) \vdash s\,]$
; *development*      $:= \langle$ Implicit development of mss (not checked). 6.6.5 $\rangle$

This code is used in section 6.6.

**6.6.5.**   The development itself is presented in forward direction by equational reasoning, similar to the original presentation in the literature. In fact, Bird's original development which appeared in [14] is reproduced in Fig. 15. Comparing that rigorous but still informal development with the formal, machine checked, Deva development reveals that there is actually not that much difference between the two. The "hints" of Bird which refer to the rule(s) that justify each transformation step translate into the "application" of those rules. The progress of the development is indicated by the judgements. It is this close (syntactic) similarity between the calculations in in the Squiggol calculus and their counterpart in Deva which gives us reason to claim that we have indeed faithfully modeled the Squiggol calculus.

**6.6.6.**   The reader will have noticed that this development is quite implicit, i.e., a lot of details are left unstated. The development becomes a bit more explicit if the two loops are unfolded into two single steps each.

$\langle$ First explicitation of the development of mss (not checked). 6.6.6 $\rangle \equiv$
*frefl*
$\quad \therefore mss = ((\uparrow) /) \circ (((+) /)*) \circ ((\mathbin{+\!\!\!+}) /) \circ (tails*) \circ inits$
$\backslash\ funfold(map\_promotion)$
$\quad \therefore mss = ((\uparrow) /) \circ ((\mathbin{+\!\!\!+}) /) \circ ((((+) /)*)*) \circ (tails*) \circ inits$
$\backslash\ funfold(reduce\_promotion(max\_props))$
$\quad \therefore mss = ((\uparrow) /) \circ (((\uparrow) /)*) \circ ((((+) /)*)*) \circ (tails*) \circ inits$
$\backslash\ funfold(map\_distribution)$
$\quad \therefore mss = ((\uparrow) /) \circ (((((\uparrow) /) \circ (((+) /)*))*) \circ (tails*) \circ inits$
$\backslash\ funfold(map\_distribution)$
$\quad \therefore mss = ((\uparrow) /) \circ (((((\uparrow) /) \circ (((+) /)*) \circ tails)*) \circ inits$
$\backslash\ funfold(horner\_rule(max\_plus\_props))$
$\quad \therefore mss = ((\uparrow) /) \circ ([a, b : s \vdash (a + b) \uparrow 0] \mathbin{\#}_0)$

**6.6.7.**   In this presentation, enough details were available to trace the transformations step by step. However, the unfold rule is used without explicitly stating the actual variant used in each application, i.e., in this example the variant *funfold* is always used (c.f. Chap. 4.4.1). Similarly, no explicit substitutions are given for the implicit parameters of *unfold*. For example, its implicitly defined functional parameter $F$ describes the position at which the law of the unfold

$$mss$$
$$= \quad \{ \text{ definition } \}$$
$$\uparrow / \circ (+/)* \circ segs$$
$$= \quad \{ \text{ definition of } segs \}$$
$$\uparrow / \circ (+/)* \circ +\!\!+/ \circ tails* \circ inits$$
$$= \quad \{ \text{ map and reduce promotion } \}$$
$$\uparrow / \circ (\uparrow / \circ (+/)* \circ tails)* \circ inits$$
$$= \quad \{ \text{ Horner's rule with } a \circledast b = (a + b) \uparrow 0 \}$$
$$\uparrow / \circ \circledast \!\!\not\!\to_0 * \circ inits$$
$$= \quad \{ \text{ accumulation lemma } \}$$
$$\uparrow / \circ \circledast \!\!\not\!\!\not\!\to_0$$

---

⟨ Implicit development of mss (not checked). 6.6.5 ⟩ ≡

*frefl*

$$\therefore mss = ((\uparrow) /) \circ (((+) /)*) \circ segs$$

— definition of *segs*

$$\therefore mss = ((\uparrow) /) \circ (((+) /)*) \circ ((+\!\!+) /) \circ (tails*) \circ inits$$

$\backslash$ **loop** *promotion_tac* (*max_props*)

$$\therefore mss = ((\uparrow) /) \circ (((\uparrow) /)*) \circ ((((+) /)*)*) \circ (tails*) \circ inits$$

$\backslash$ **loop** *unfold* (*map_distribution*)

$$\therefore mss = ((\uparrow) /) \circ ((((\uparrow) /) \circ (((+) /)*) \circ tails)*) \circ inits$$

$\backslash$ *unfold*(*horner_rule*(*max_plus_props*))

$$\therefore mss = ((\uparrow) /) \circ (([a, b : s \vdash (a + b) \uparrow 0]\!\!\not\!\to_0)*) \circ inits$$

— definition of left accumulation

$$\therefore= mss = ((\uparrow) /) \circ ([a, b : s \vdash (a + b) \uparrow 0]\!\!\not\!\!\not\!\to_0)$$

This code is used in section 6.6.4.

**Fig. 15.** The classic development of the maximum segment sum algorithm and its Deva formalization

rule is applied. When specifying the precise variant of *unfold* used and adding the substitutions for the parameter $F$, the development looks as follows.

⟨ Second explicitation of the development of mss. 6.6.7 ⟩ ≡

*frefl*

$$\therefore mss = ((\uparrow) /) \circ (((+) /)*) \circ ((+\!\!+) /) \circ (tails*) \circ inits$$

$\backslash$ *funfold*($F := [f : [ list (list(list(s))) \vdash list(s)] \vdash ((\uparrow) /) \circ f \circ (tails*) \circ inits ],$
$\qquad map\_promotion$)

$\therefore mss = ((\uparrow) /) \circ ((+\!\!+) /) \circ ((((+) /)*)*) \circ (tails*) \circ inits$
$\setminus funfold(F := [\, f :[\, list\, (list(s)) \vdash s\,] \vdash f \circ ((((+) /)*)*) \circ (tails*) \circ inits\,],$
$\qquad\qquad reduce\_promotion\ (max\_props))$
$\therefore mss = ((\uparrow) /) \circ (((\uparrow) /)*) \circ ((((+) /)*)*) \circ (tails*) \circ inits$
$\setminus funfold(F := [\, f :[\, list\, (list(list(s))) \vdash list(s)\,] \vdash ((\uparrow) /) \circ f \circ (tails*) \circ inits\,],$
$\qquad\qquad map\_distribution)$
$\therefore mss = ((\uparrow) /) \circ ((((\uparrow) /) \circ (((+) /.)*))*) \circ (tails*) \circ inits$
$\setminus funfold(F := [\, f :[\, list\, (list(s)) \vdash list(s)\,] \vdash ((\uparrow) /) \circ f \circ inits\,],$
$\qquad\qquad map\_distribution)$
$\therefore mss = ((\uparrow) /) \circ ((((\uparrow) /) \circ (((+) /)*) \circ tails)*) \circ inits$
$\setminus funfold(F := [\, f :[\, list\, (s) \vdash s\,] \vdash ((\uparrow) /) \circ (f*) \circ inits\,],$
$\qquad\qquad horner\_rule\ (max\_plus\_props))$
$\therefore mss = ((\uparrow) /) \circ ([a, b :\ s\ \vdash (a + b) \uparrow 0]\!\!\not\!\#_0)$

**6.6.8.** The preceding version is still far from being fully explicit, there are many implicitly defined parameters of the unfold rule and the involved laws that are not explicitly given. After inserting explicit substitutions for these parameters, the development looks as follows.

$\langle$ Third explicitation of the development of mss. 6.6.8 $\rangle \equiv$
*frefl*
$\quad \therefore mss = ((\uparrow) /) \circ (((+) /)*) \circ ((+\!\!+) /) \circ (tails*) \circ inits$
$\setminus funfold(f := (((+) /)*) \circ ((+\!\!+) /), g := ((+\!\!+) /) \circ ((((+) /)*)*),$
$\qquad\qquad F := [\, f :[\, list\, (list(list(s))) \vdash list(s)\,] \vdash ((\uparrow) /) \circ f \circ (tails*) \circ inits\,],$
$\qquad\qquad map\_promotion\ (f := ((+) /)))$
$\quad \therefore mss = ((\uparrow) /) \circ ((+\!\!+) /) \circ ((((+) /)*)*) \circ (tails*) \circ inits$
$\setminus funfold(f := ((\uparrow) /) \circ ((+\!\!+) /), g := ((\uparrow) /) \circ (((\uparrow) /)*),$
$\qquad\qquad F := [\, f :[\, list\, (list(s)) \vdash s\,] \vdash f \circ ((((+) /)*)*) \circ (tails*) \circ inits\,],$
$\qquad\qquad reduce\_promotion\ (\oplus := (\uparrow), 1_\oplus := \bot, max\_props))$
$\quad \therefore mss = ((\uparrow) /) \circ (((\uparrow) /)*) \circ ((((+) /)*)*) \circ (tails*) \circ inits$
$\setminus funfold(f := (((\uparrow) /)*) \circ ((((+) /)*)*), g := (((\uparrow) /) \circ (((+) /)*))*,$
$\qquad\qquad F := [\, f :[\, list\, (list(list(s))) \vdash list(s)\,] \vdash ((\uparrow) /) \circ f \circ (tails*) \circ inits\,],$
$\qquad\qquad map\_distribution\ (f := ((\uparrow) /), g := ((+) /)*))$
$\quad \therefore mss = ((\uparrow) /) \circ ((((\uparrow) /) \circ (((+) /)*))*) \circ (tails*) \circ inits$
$\setminus funfold(f := ((((\uparrow) /) \circ (((+) /)*))*) \circ (tails*),$
$\qquad\qquad g := (((\uparrow) /) \circ (((+) /)*) \circ tails)*,$
$\qquad\qquad F := [\, f :[\, list\, (list(s)) \vdash list(s)\,] \vdash ((\uparrow) /) \circ f \circ inits\,],$
$\qquad\qquad map\_distribution\ (f := ((\uparrow) /) \circ (((+) /)*), g := tails))$

$$\therefore mss = ((\uparrow) \, /) \circ ((((\uparrow) \, /) \circ (((+) \, /)*) \circ \mathit{tails})*) \circ \mathit{inits}$$
$$\backslash \mathit{funfold}(f := ((\uparrow) \, /) \circ (((+) \, /)*) \circ \mathit{tails}, g := [a, b : s \vdash (a + b) \uparrow 0] \not\to_0,$$
$$\qquad F := [f : [\mathit{list}(s) \vdash s] \vdash ((\uparrow) \, /) \circ (f*) \circ \mathit{inits}],$$
$$\qquad \mathit{horner\_rule} \, (\oplus := (\uparrow), \otimes := (+), 1_\oplus := \bot, 1_\otimes := 0, \mathit{max\_plus\_props}))$$
$$\therefore mss = ((\uparrow) \, /) \circ ([a, b : s \vdash (a + b) \uparrow 0] \not\to_0)$$

This Deva text looks quite horrible, yet it is still far from being completely explicit: There remain those implicit arguments which instantiate the constructions to the proper sorts. Their number is vast, e.g. every use of / and * has two such implicit arguments. A fully explicit version of the development is not presented, it would be an explosion wrt. the first development.

These consecutive *explicitations* of the maximum segment sum development have been shown in order to motivate the need for suppressing annoying details of developments. The developer should be allowed to think only in terms of the implicit development or its first explicitation. The second and third explicitations, and even further ones, should be consulted only occasionally to check things such as instantiation details.

## 6.7  Tree Evaluation Problems

The second development presented in this chapter concerns the $\alpha\beta$-pruning strategy for two-player game-trees.

⟨ Tree evaluation problems. 6.7 ⟩ ≡
**context**  *TreeEvaluationProblems* :=
〚 **import**   *TheoryOfNonEmptyJoinLists*
; ⟨ Arbitrary branching trees. 6.7.1 ⟩
; ⟨ Minimax evaluation. 6.7.2 ⟩
; ⟨ Interval windows. 6.7.3 ⟩
; ⟨ $\alpha\beta$-pruning. 6.7.13 ⟩
〛

This code is used in section 6.1.

**6.7.1.**   The context below defines the constructors for non-empty trees with an arbitrary but finite branching degree. Trees have values at their tips and a non-empty list of subtrees at their forks. It is possible to introduce the second-order operators on trees which correspond to those on lists (cf. [79]).

⟨ Arbitrary branching trees. 6.7.1 ⟩ ≡
   *tree*  : [*sort* ⊢ *sort*]
; *tip*   : [ *s* ? *sort* ; *s* ⊢ *tree*(*s*)]

; *fork* : [ *s* ? *sort* ; *ne_list*(*tree*(*s*)) ⊢ *tree*(*s*) ]

This code is used in section 6.7.

**6.7.2.** A *game-tree* is a tree that describes the possible moves of a two-player game up to some given depth. The *minimax* evaluation of a game tree starts with a rating given for all the leaf positions of the tree. It then computes the ratings of the moves further up in the tree in a bottom-up, minimax fashion. The game-tree in Figure 16 illustrates the minimax evaluation scheme. A circular node denotes a move of the player whereas a square node denotes a move of her opponent. The tree shows two possible moves of the player which have both two possible reactions of the opponent. Further levels are not shown. The numbers (ranging over the set of integers) denote the rating associated with each move: At a circular node they are a lower bound for the value to be reached by the player at that move, whatever the opponents will do. Conversely, at a square node they are an upper bound for the value to be reached by the opponent at that move, whatever the player will do. The numbers can be computed bottom up by the so-called *minimax-evaluation*: i.e., the rating of a leaf is given and left unchanged. The rating of a circular node results from taking the maximum of the ratings associated with its immediate descendant square nodes while the rating of a square node results from taking the minimum of its immediate descendant circular nodes.

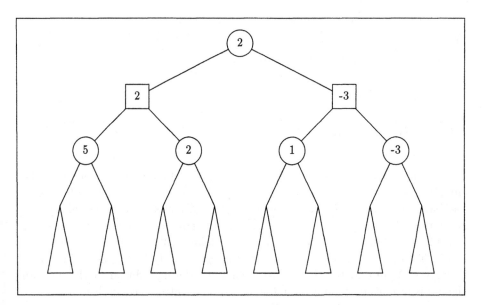

**Fig. 16.** Minimax evaluation

Minimax evaluation *eval*(*t*) of a tree is recursively defined below: While on tip positions the evaluation just returns the value associated with the tip, on fork positions all branches are recursively evaluated and the results of this evaluation, which correspond to the players (or the opponents) optimization, are negated and their maximum is computed, i.e.,

$$
\begin{aligned}
eval(tip(n)) &= n \\
eval(fork(tl)) &= \uparrow /((- \circ eval) * tl) \\
&\qquad \{\text{notation for list catamorphisms}\} \\
&= (\!|\uparrow, - \circ eval|\!)(tl)
\end{aligned}
$$

Equivalently, the minimum of the results could be computed and then negated. Because of the double negation law $-(-a) = a$, the evaluation successively minimizes and maximizes on each level up.

The formalization in Deva abstracts from the concrete type of integers as used in Fig. 16. Instead, an arbitrary sort $s$ of values is considered which we assume to be equipped with the structure of a boolean algebra.

$\langle$ Minimax evaluation. 6.7.2 $\rangle \equiv$

| | |
|---|---|
| $s$ | : $sort$ |
| $; \perp, \top$ | : $s$ |
| $; -(\cdot)$ | : $[s \vdash s]$ |
| $; (\cdot) \uparrow (\cdot), (\cdot) \downarrow (\cdot)$ | : $[s; s \vdash s]$ |
| $; ba$ | : $boolean\_algebra \ (\perp, \top, (\uparrow), (\downarrow), -)$ |
| $; eval$ | : $[\, tree \ (s) \vdash s \,]$ |
| $; eval\_def$ | : $\langle\ tips \ := eval \circ tip = id$ |
| | $, forks := eval \circ fork = (\!|(\uparrow), (-) \circ eval|\!)$ |
| | $\rangle$ |

This code is used in section 6.7.

**6.7.3.** The idea underlying $\alpha\beta$-*pruning* is to record intervals within which the minimax value of the node may fall. Once one knows for sure that the rating associated with an immediate descendant node falls outside this interval, further evaluation of the subtree (of which the descendant node is the root) may be discontinued. The context we are about to define will introduces the function $I_{\alpha}^{\beta}$ which "coerces" its argument into the interval $[\alpha, \beta]$. First, the definition of $I_{\alpha}^{\beta}$ is given: Note that the theory of the partial ordering induced by the boolean algebra is imported, making available all the laws introduced in Sect. 4.4.3.

$\langle$ Interval windows. 6.7.3 $\rangle \equiv$

**import** $PartialOrdering \ (ba)$

$; I_{(\cdot)}^{(\cdot)} := [\alpha, \beta : s \vdash (\beta \downarrow) \circ (\alpha \uparrow)]$

; ⟨ Properties of interval windows. 6.7.5 ⟩

This code is used in section 6.7.

**6.7.4.** Note that the definition of $I_\alpha^\beta$ is too loose, since it does not enforce the constraint that $\alpha \sqsubseteq \beta$. Alternatively the constraint could be introduced directly as a condition of the coercion function, however this would cause notational clutter. With the given definition, the constraint will come up anyway as a condition in a lot of the desired properties of interval windows. The first such property is that if the interval is $[\bot, \top]$ then the interval window is the identity function. The derivation of this property consists of a straightforward boolean simplification, details of which are suppressed by using the tactic *bool_simp* repeatedly. As illustrated in the derivation of the maximum segment sum, the iteration can be unfolded into individual boolean transformation steps.

**6.7.5.**      ⟨ Properties of interval windows. 6.7.5 ⟩ ≡

$$I\_bottop := [\, a \qquad ? \quad s$$
$$; \ LHS := I_\bot^\top(a)$$
$$\vdash trefl$$
$$\therefore LHS = \top \downarrow (\bot \uparrow a)$$
$$\backslash \ \textbf{loop} \ bool\_simp \ (ba)$$
$$\therefore LHS = a$$
$$]$$
$$\backslash \ extensionality. \ down$$
$$\therefore I_\bot^\top = id$$

See also sections 6.7.6, 6.7.7, 6.7.8, 6.7.9, 6.7.10, and 6.7.11.

This code is used in section 6.7.3.

**6.7.6.** The next property states that the interval window function always delivers results smaller than its upper bound. The proof consists of a single application of the absorption law.

⟨ Properties of interval windows. 6.7.5 ⟩+ ≡
$$; \ I\_zero := [\alpha, \beta, a \ ? \quad s$$
$$; \ LHS := \beta \uparrow I_\alpha^\beta(a)$$
$$\vdash trefl$$
$$\therefore LHS = \beta \uparrow (\beta \downarrow (\alpha \uparrow a))$$
$$\backslash \ unfold(ba. \ absorp \ .join)$$
$$\therefore LHS = \beta$$
$$]$$
$$\therefore [\alpha, \beta, a \ ? \ s \vdash \beta \uparrow I_\alpha^\beta(a) = \beta]$$

**6.7.7.** An interval window $I_\alpha^\beta$ is narrowed by post-composing it with $(a \uparrow)$, for some value $a$ with $a \sqsubseteq \beta$. The proof is a slightly more complex boolean transformation; two transformation steps depend upon the condition $a \sqsubseteq \beta$, one step applies the associativity law.

$\langle$ Properties of interval windows. 6.7.5 $\rangle+ \equiv$
; $I\_narrow := [\alpha, \beta, a, b \;?\; s \;;\; hyp : \langle\alpha \sqsubseteq a, a \sqsubseteq \beta\rangle; \; LHS := a \uparrow I_\alpha^\beta(b)$
$\qquad \vdash trefl$
$\qquad\qquad \therefore LHS = a \uparrow (\beta \downarrow (\alpha \uparrow b))$
$\qquad\qquad \backslash fold(exch\_meet\_join(hyp.\mathbf{2}))$
$\qquad\qquad \therefore LHS = \beta \downarrow (a \uparrow (\alpha \uparrow b))$
$\qquad\qquad \backslash fold(ba. \; assoc \; .join)$
$\qquad\qquad \therefore LHS = \beta \downarrow ((a \uparrow \alpha) \uparrow b)$
$\qquad\qquad \backslash fold(ba. \; commut \; .join)$
$\qquad\qquad \therefore LHS = \beta \downarrow ((\alpha \uparrow a) \uparrow b)$
$\qquad\qquad \backslash unfold(hyp.\mathbf{1})$
$\qquad\qquad \therefore LHS = \beta \downarrow (a \uparrow b)$
$\qquad\qquad \therefore LHS = I_a^\beta(b)$
$\quad]$

$$\therefore [\alpha, \beta, a, b \;?\; s \; \vdash \frac{\langle\alpha \sqsubseteq a, a \sqsubseteq \beta\rangle}{a \uparrow (I_\alpha^\beta(b)) = I_a^\beta(b)}]$$

**6.7.8.** *I_narrow* can be specialized to the case of narrowing an interval window with its own lower bound.

$\langle$ Properties of interval windows. 6.7.5 $\rangle+ \equiv$
; $I\_one := [\alpha, \beta, a \;?\; s \;;\; hyp : \alpha \sqsubseteq \beta$
$\qquad \vdash I\_narrow \; (\langle refl\_smth \therefore \alpha \sqsubseteq \alpha, hyp \rangle)$
$\qquad\qquad \therefore \alpha \uparrow I_\alpha^\beta(a) = I_\alpha^\beta(a)$
$\quad]$

$$\therefore [\alpha, \beta, a \;?\; s \; \vdash \frac{\alpha \sqsubseteq \beta}{\alpha \uparrow I_\alpha^\beta(a) = I_\alpha^\beta(a)}]$$

**6.7.9.** Applying an interval window to a negated value is equivalent to negating and exchanging the bounds of the interval window, applying it, and negating the result. The proof, which makes use of the tactic moving a negation inside a boolean expression, can be presented more elegantly when starting from the right-hand side of the desired equation.

⟨ Properties of interval windows. 6.7.5 ⟩+ ≡

; $I\_neg\_shift := [\alpha, \beta\ ?\ s\ ;\ hyp\ :\ \alpha \sqsubseteq \beta$

$\qquad \vdash [\ a\ ?\ s\ ;\ RHS := -\ I_{-\beta}^{-\alpha}(a)$

$\qquad\quad \vdash trefl$

$\qquad\qquad \therefore RHS = -(-\alpha \downarrow (-\beta \uparrow a))$

$\qquad\qquad\quad \backslash\ \textbf{loop}\ bool\_neg\_simp\ (ba)$

$\qquad\qquad\qquad \therefore RHS = \alpha \uparrow (\beta \downarrow -a)$

$\qquad\qquad\quad \backslash\ fold(exch\_meet\_join(hyp))$

$\qquad\qquad\qquad \therefore RHS = \beta \downarrow (\alpha \uparrow -a)$

$\qquad\qquad\qquad \therefore RHS = I_{\alpha}^{\beta}(-a)$

$\qquad\qquad\quad \backslash\ sym$

$\qquad\qquad\qquad \therefore I_{\alpha}^{\beta}(-a) = RHS$

$\qquad\quad ]$

$\qquad\quad \backslash\ extensionality.\ down$

$\qquad ]$

$\qquad \therefore [\alpha, \beta\ ?\ s\ \vdash \dfrac{\alpha \sqsubseteq \beta}{I_{\alpha}^{\beta} \circ (-) = (-) \circ I_{-\beta}^{-\alpha}} ]$

**6.7.10.** Interval windows commute with ↑. The proof involves reflexivity, some associative-commutative manipulations, and finally a distributivity law.

⟨ Properties of interval windows. 6.7.5 ⟩+ ≡

; $I\_max\_shift := [\alpha, \beta, a, b\ ?\ s\ ;\ LHS := I_{\alpha}^{\beta}(a \uparrow b)$

$\qquad\quad \vdash trefl$

$\qquad\qquad \therefore LHS = \beta \downarrow (\alpha \uparrow (a \uparrow b))$

$\qquad\qquad\quad \backslash\ fold(refl\_smth)$

$\qquad\qquad\qquad \therefore LHS = \beta \downarrow ((\alpha \uparrow \alpha) \uparrow (a \uparrow b))$

$\qquad\qquad\quad \backslash\ \textbf{loop}\ bool\_ac\ (ba)$

$\qquad\qquad\qquad \therefore LHS = \beta \downarrow ((\alpha \uparrow a) \uparrow (\alpha \uparrow b))$

$\qquad\qquad\quad \backslash\ unfold(ba.\ distrib\ .meet)$

$\qquad\qquad\qquad \therefore LHS = (\beta \downarrow (\alpha \uparrow a)) \uparrow (\beta \downarrow (\alpha \uparrow b))$

$\qquad\qquad\qquad \therefore LHS = I_{\alpha}^{\beta}(a) \uparrow I_{\alpha}^{\beta}(b)$

$\qquad\quad ]$

$\qquad \therefore [\alpha, \beta, a, b\ ?\ s\ \vdash I_{\alpha}^{\beta}(a \uparrow b) = I_{\alpha}^{\beta}(a) \uparrow I_{\alpha}^{\beta}(b)]$

**6.7.11.** *I_max_shift* is now used to resolve the condition of the promotion theorem (cf. Sect. 6.4.3), when proving that interval windows can be promoted within catamorphisms of the form $([\uparrow, f])$.

⟨Properties of interval windows. 6.7.5⟩+ ≡

; *I_hom_promote* :=

[ *t* ? *sort* ; α, β ? *s* ; *f* ? [ *t* ⊢ *s* ]; *LHS* := $I_\alpha^\beta \circ (\!(\uparrow), f)\!)$

⊢ *frefl*

    ∴ *LHS* = $I_\alpha^\beta \circ ((\uparrow) /) \circ (f*)$

    \ *unfold*(*promotion*(⟨ *ba* . *assoc* . *join* , *ba* . *assoc* . *join* ⟩, $I_\alpha^\beta$).*down*)(*I_max_shift*)

    ∴ *LHS* = $((\uparrow) /) \circ (I_\alpha^\beta *) \circ (f*)$

    \ *unfold*(*map_distribution*)

    ∴ *LHS* = $((\uparrow) /) \circ ((I_\alpha^\beta \circ f)*)$

    ∴ *LHS* = $(\!(\uparrow), I_\alpha^\beta \circ f)\!)$

]

    ∴ [ *t* ? *sort* ; α, β ? *s* ; *f* ? [ *t* ⊢ *s* ] ⊢ $I_\alpha^\beta \circ (\!(\uparrow), f)\!) = (\!(\uparrow), I_\alpha^\beta \circ f)\!)$ ]


**6.7.12.**    The αβ-algorithm prunes the game tree by computing boundaries [α,β] within which the minimax value of a node will fall: For example, assume that one player has already evaluated one of her possible moves to, say, a rating of α. Now, take any of her other possible moves (call it *A*). Assume that her opponent can respond to *A* by a move *B* rated *x* such that *x* ⊑ α, i.e., a move worse than α from her point of view. The αβ-pruning strategy now allows to conclude that there is no purpose in evaluating any further reactions to *A*, because the opponent is assumed to play optimal and thus will choose a reaction to *A* which is at least as good (from his view) as *B*. This strategy can be illustrated on the game-tree shown in Fig. 17. The left subtree corresponds to one of the two possible moves of the player, and yields at least a value of 2, whatever action the opponent chooses. Now, consider the move corresponding to the right subtree: Since the opponent's left reaction already yields a value of 1, it is unnecessary to inspect its right reaction since the player's second move will never obtain a value greater than 1.

    This strategy requires some bookkeeping in the algorithm, which is achieved here in form of intervals in which the minimax values must fall in order to prevent pruning. Further, it requires some form of direction of the evaluation, which is a left-order tree traversal in this presentation. In Figure 17, the tree is decorated with such intervals, computed by a left order traversal. If there is a node on the level below an interval whose value does not fit into the interval, the tree can be pruned. Since the value 1 of the third node on the bottom level does not fall within the previously computed interval [2, ⊤], the subtree belonging to the node marked with * can be cut off.

    The development of the algorithm proceeds in several steps: First, a bounded variant (leval$_\alpha^\beta$) of the minimax evaluation schemed can be defined by coercing the values of each player node into the given interval [α, β].


**6.7.13.**    ⟨αβ-pruning. 6.7.13⟩ ≡

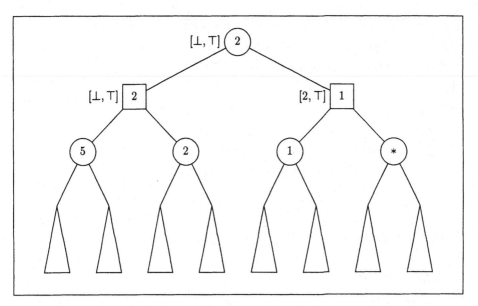

**Fig. 17.** $\alpha\beta$-pruning

$$\mathsf{leval}_{(\cdot)}^{(\cdot)} := [\alpha, \beta : s \vdash \mathsf{l}_\alpha^\beta \circ \mathit{eval}]$$

See also sections 6.7.14, 6.7.15, 6.7.18, 6.7.19, 6.7.20, 6.7.21, 6.7.22, and 6.7.24.

This code is used in section 6.7.

**6.7.14.** Using an elementary property of interval windows, it can then be shown that $\mathsf{leval}_\bot^\top$ is equivalent to the minimax evaluation scheme.

$\langle\,\alpha\beta\text{-pruning. 6.7.13}\,\rangle + \equiv$
; *correctness* := *frefl*
$$\therefore \mathsf{leval}_\bot^\top = \mathsf{l}_\bot^\top \circ \mathit{eval}$$
$$\backslash\ \mathit{unfold}(I\_bottop)$$
$$\therefore \mathsf{leval}_\bot^\top = \mathit{id} \circ \mathit{eval}$$
$$\backslash\ \mathit{unfold}(\mathit{pid}.\mathit{left})$$
$$\therefore \mathsf{leval}_\bot^\top = \mathit{eval}$$

**6.7.15.** A recursive characterization of $\mathsf{leval}_\alpha^\beta$ can be derived by a simple development.

$\langle\,\alpha\beta\text{-pruning. 6.7.13}\,\rangle + \equiv$
; *ieval_rec* :=

$[\alpha, \beta\ ?\ s\ ;\ hyp\ :\alpha \sqsubseteq \beta;\ LHS_1 := \mathsf{leval}^{\beta}_{\alpha} \circ tip;\ LHS_2 := \mathsf{leval}^{\beta}_{\alpha} \circ fork$
$\vdash\!($ $tips\ := \langle$ Calculation of $\mathsf{leval}^{\beta}_{\alpha} \circ tip.\ 6.7.16 \rangle$
$,\ forks := \langle$ Calculation of $\mathsf{leval}^{\beta}_{\alpha} \circ fork.\ 6.7.17 \rangle$
$)$
$]$

$$\therefore [\alpha, \beta\ ?\ s\ \vdash \left|\begin{array}{l} \underline{\qquad\qquad\alpha \sqsubseteq \beta \qquad\qquad} \\ (\ tips\ := \mathsf{leval}^{\beta}_{\alpha} \circ tip = \mathsf{I}^{\beta}_{\alpha} \\ ,\ forks := \mathsf{leval}^{\beta}_{\alpha} \circ fork = (\!(\uparrow), (-) \circ \mathsf{leval}^{-\alpha}_{-\beta}]\!)\ ] \\ ) \end{array}\right.$$

**6.7.16.** $\langle$ Calculation of $\mathsf{leval}^{\beta}_{\alpha} \circ tip.\ 6.7.16 \rangle \equiv$
*frefl*
  $\therefore LHS_1 = \mathsf{I}^{\beta}_{\alpha} \circ eval \circ tip$
$\backslash\ unfold(eval\_def.tips)$
  $\therefore LHS_1 = \mathsf{I}^{\beta}_{\alpha} \circ id$
$\backslash\ unfold(pid.right)$
  $\therefore LHS_1 = \mathsf{I}^{\beta}_{\alpha}$

This code is used in section 6.7.15.

**6.7.17.** $\langle$ Calculation of $\mathsf{leval}^{\beta}_{\alpha} \circ fork.\ 6.7.17 \rangle \equiv$
*frefl*
  $\therefore LHS_2 = \mathsf{I}^{\beta}_{\alpha} \circ eval \circ fork$
$\backslash\ unfold(eval\_def.forks)$
  $\therefore LHS_2 = \mathsf{I}^{\beta}_{\alpha} \circ (\!(\uparrow), ((-)\!) \circ eval))$
$\backslash\ unfold(I\_hom\_promote)$
  $\therefore LHS_2 = (\!(\uparrow), (\mathsf{I}^{\beta}_{\alpha} \circ (-) \circ eval)]\!)$
$\backslash\ unfold(I\_neg\_shift(hyp))$
  $\therefore LHS_2 = (\!(\uparrow), (-) \circ \mathsf{I}^{-\alpha}_{-\beta} \circ eval]\!)$
  $\therefore LHS_2 = (\!(\uparrow), (-) \circ \mathsf{leval}^{-\alpha}_{-\beta}]\!)$

This code is used in section 6.7.15.

**6.7.18.** In the second step, the (currently undirected) evaluation of the subtrees will be specialized into a directed evaluation from left to right. In calculational algorithm calculation, this design step corresponds to specializing the catamorphism $(\!(\uparrow, - \circ \mathsf{leval}^{-\alpha}_{-\beta}]\!)$ derived above into a left-reduction. This specialization will introduce an auxiliary operator *next*, which updates the current minimax value $a$ by consideration of the next subtree $t$. *next* is defined as follows:

$\langle \alpha\beta\text{-pruning. } 6.7.13 \rangle + \equiv$

$; next := [\alpha, \beta : s ; a : s ; t : tree\,(s) \vdash a \uparrow -(\text{leval}_{-\beta}^{-\alpha}(t))]$

**6.7.19.** In order to apply the specialization law of non-empty lists to a catamorphism $(\!(\uparrow, f)\!)$ (cf. Sect. 6.5), a left-identity of $\uparrow$ on the range of $f$ must be
found. The lemma below establishes $\alpha$ as a left-identity of $\uparrow$ on the range of
$- \circ \text{leval}_{-\beta}^{-\beta}$.

$\langle \alpha\beta\text{-pruning. } 6.7.13 \rangle + \equiv$

$; spec\_cond := [\alpha, \beta, b\,?\,\,s ; hyp : \alpha \sqsubseteq \beta; LHS := \alpha \uparrow (-(\mathsf{l}_{-\beta}^{-\alpha}(b)))$

$\qquad\qquad\qquad \vdash trefl$

$\qquad\qquad\qquad\quad \therefore LHS = \alpha \uparrow ((-) \circ \mathsf{l}_{-\beta}^{-\alpha})(b)$

$\qquad\qquad\qquad\quad \backslash\ fold(I\_neg\_shift(hyp))$

$\qquad\qquad\qquad\quad \therefore LHS = \alpha \uparrow \mathsf{l}_{\alpha}^{\beta}(-b)$

$\qquad\qquad\qquad\quad \backslash\ unfold(I\_one(hyp))$

$\qquad\qquad\qquad\quad \therefore LHS = (\mathsf{l}_{\alpha}^{\beta} \circ (-))(b)$

$\qquad\qquad\qquad\quad \backslash\ unfold(I\_neg\_shift(hyp))$

$\qquad\qquad\qquad\quad \therefore LHS = -(\mathsf{l}_{-\beta}^{-\alpha}(b))$

$\qquad\qquad ]$

$$\therefore [\alpha, \beta, b\,?\,\,s \vdash \dfrac{\alpha \sqsubseteq \beta}{\alpha \uparrow - \mathsf{l}_{-\beta}^{-\alpha}(b) = -\mathsf{l}_{-\beta}^{-\alpha}(b)}]$$

**6.7.20.** A straightforward application of specialization now yields the desired
left-reduction.

$\langle \alpha\beta\text{-pruning. } 6.7.13 \rangle + \equiv$

$; specialize :=$

$[\alpha, \beta\,?\,\,s ; hyp : \alpha \sqsubseteq \beta; LHS := \text{leval}_{\alpha}^{\beta} \circ fork$

$\vdash [\,lemma := spec\_cond\,(hyp)$

$\qquad\qquad\quad \therefore [\,t\,?\,tree\,(s) \vdash \alpha \uparrow - \mathsf{l}_{-\beta}^{-\alpha}(eval(t)) = -\mathsf{l}_{-\beta}^{-\alpha}(eval(t))]$

$\quad \vdash ieval\_rec\,(hyp).\,forks$

$\qquad\quad \therefore LHS = (\!(\,(\uparrow), (-) \circ \text{leval}_{-\beta}^{-\alpha}\,)\!)$

$\qquad\quad \backslash\ unfold(specialization(ba.\ assoc\,.join, lemma).left)$

$\qquad\quad \therefore LHS = next(\alpha, \beta) \not\nearrow_{\alpha}$

$\quad ]$

$]$

$$\therefore [\alpha, \beta\,?\,\,s \vdash \dfrac{\alpha \sqsubseteq \beta}{\text{leval}_{\alpha}^{\beta} \circ fork = next(\alpha, \beta) \not\nearrow_{\alpha}}]$$

**6.7.21.** In the third and crucial step, two properties about the operator *next* are derived which will allow the actual pruning of game-trees: The first property is the fact that the evaluation of the moves already considered may lift the lower bound for the next move. The proof is quite analogous to the above proof of the specialization condition, except that this time the rule *I_narrow* is used.

$\langle \alpha\beta\text{-pruning. } 6.7.13 \rangle + \equiv$
$; \; lift := [\alpha, \beta, a \; ? \; s \; ; \; t \; ? \; tree\,(s)$
$\qquad ; \; hyp \qquad : \quad \langle\!\langle \alpha \sqsubseteq a, a \sqsubseteq \beta \rangle\!\rangle$
$\qquad ; \; lemma := trans\_smth\,(hyp.1, hyp.2) \therefore \alpha \sqsubseteq \beta$
$\qquad ; \; LHS \quad := \; next\,(\alpha, \beta, a, t)$
$\qquad \vdash trefl$
$\qquad\qquad \therefore LHS = a \uparrow ((-) \circ \mathsf{I}^{-\alpha}_{-\beta})(eval(t))$
$\qquad\qquad \backslash \; fold\,(I\_neg\_shift\,(lemma))$
$\qquad\qquad \therefore LHS = a \uparrow \mathsf{I}^{\beta}_{\alpha}(-(eval(t)))$
$\qquad\qquad \backslash \; unfold\,(I\_narrow\,(hyp))$
$\qquad\qquad \therefore LHS = (\mathsf{I}^{\beta}_{a} \circ (-))(eval(t))$
$\qquad\qquad \backslash \; unfold\,(I\_neg\_shift\,(hyp.2))$
$\qquad\qquad \therefore LHS = -(\mathsf{leval}^{-a}_{-\beta}(t))$
$\qquad ]$

$$\therefore [\alpha, \beta, a \; ? \; s \; ; \; t \; ? \; tree\,(s) \vdash \frac{\langle\!\langle \alpha \sqsubseteq a, a \sqsubseteq \beta \rangle\!\rangle}{next\,(\alpha, \beta, a, t) = -(\mathsf{leval}^{-a}_{-\beta}(t))}]$$

**6.7.22.** The second property of *next* is the fact that in case the evaluation ever reaches the upper bound, the remaining moves can be pruned. The derivation consists essentially of applying the law *I_zero*.

$\langle \alpha\beta\text{-pruning. } 6.7.13 \rangle + \equiv$
$; \; prune := [\alpha, \beta \; ? \; s \; ; \; t \; ? \; tree\,(s); \; hyp : \alpha \sqsubseteq \beta; \; LHS := next\,(\alpha, \beta, \beta, t)$
$\qquad\qquad \vdash trefl$
$\qquad\qquad\qquad \therefore LHS = \beta \uparrow ((-) \circ \mathsf{I}^{-\alpha}_{-\beta})(eval(t))$
$\qquad\qquad \backslash \; fold\,(I\_neg\_shift\,(hyp))$
$\qquad\qquad\qquad \therefore LHS = \beta \uparrow \mathsf{I}^{\beta}_{\alpha}(- \, eval\,(t))$
$\qquad\qquad \backslash \; unfold\,(I\_zero)$
$\qquad\qquad\qquad \therefore LHS = \beta$
$\qquad ]$

$$\therefore [\alpha, \beta \; ? \; s \; ; \; t \; ? \; tree\,(s) \vdash \frac{\alpha \sqsubseteq \beta}{next\,(\alpha, \beta, \beta, t) = \beta}]$$

**6.7.23.** Finally, the results can be summarized into a recursive system of *conditional* equations for $\text{leval}_\alpha^\beta$.

⟨ Recursive equations for $\alpha\beta$-pruning. 6.7.23 ⟩ ≡

$$\langle\!\langle \alpha \sqsubseteq a, a \sqsubseteq \beta \rangle\!\rangle$$
$$\langle\!\langle eval = \text{leval}_\bot^\top$$
$$, \text{leval}_\alpha^\beta \circ tip = \text{I}_\alpha^\beta$$
$$, \text{leval}_\alpha^\beta \circ fork = next(\alpha, \beta) \not\rightarrow_\alpha$$
$$, [\, t \,?\, tree \,(s) \vdash next(\alpha, \beta, a, t) = -(\text{leval}_{-\beta}^{-\alpha}(t))\,]$$
$$, [\, t \,?\, tree \,(s) \vdash next(\alpha, \beta, \beta, t) = \beta\,]$$
$$\rangle\!\rangle$$

This code is used in section 6.7.24.

**6.7.24.** These equations can be seen as a functional program for $\alpha\beta$-pruning. The conditions in these equations (i.e., $\alpha \sqsubseteq a$ and $a \sqsubseteq \beta$) arise from several sources: First, although defined without any conditions, the notion of interval windows $\text{I}_\alpha^\beta$ makes only sense for those values $\alpha$ and $\beta$ for which $\alpha \sqsubseteq \beta$ holds. This is reflected by the conditions which most of the properties about interval windows require. Second, the condition that the argument of *next* must fall within the interval between $\alpha$ and $\beta$ is actually an invariant assertion of the algorithm.

⟨ $\alpha\beta$-pruning. 6.7.13 ⟩+ ≡

$$; \alpha\beta\_pruning :=$$
$$[\alpha, \beta, a \,?\, s$$
$$\vdash [\, hyp \quad : \quad \langle\!\langle \alpha \sqsubseteq a, a \sqsubseteq \beta \rangle\!\rangle$$
$$; lemma := trans\_smth\,(hyp.\mathbf{1}, hyp.\mathbf{2}) \therefore \alpha \sqsubseteq \beta$$
$$\vdash\!\langle sym\,(correctness)$$
$$, ieval\_rec\,(lemma).\,tips$$
$$, specialize\,(lemma)$$
$$, lift\,(hyp)$$
$$, prune\,(lemma)$$
$$\rangle$$
$$]$$
$$]$$
$$\therefore [\alpha, \beta, a \,?\, s \vdash \langle \text{Recursive equations for } \alpha\beta\text{-pruning. 6.7.23} \rangle]$$

## 6.8 Discussion

Currently, algorithm calculation is used predominantly as a paper-and-pencil method. In this case study we have made an attempt to completely formalize its developments and their underlying theory. We will try to evaluate this attempt in the next chapter, i.e., discuss drawbacks and benefits, also in comparison with the VDM case study. At this place we would like to briefly discuss some phenomena that are more specific to the nature of algorithm calculation.

From a syntactical point of view, we can say that the use of (sectioned) infix symbols and of super- and subscripts was sufficient to capture many common notations of algorithm calculation. On the other hand, some notational conventions exceed infix operators as e.g. the notation for finite lists (cf. Sect. 6.2).

As for all other examples, we do not make any semantic adequacy proofs in this book. In the case of algorithm calculation, this would hardly be possible anyway since there has not yet been developed any "definite" theory. Intuitively however, we claim to have captured essential aspects of the methodology.

Remember that the development of a tree pruning strategy (cf. Sect. 6.7), led to a system of conditional equations. Actually, such conditions are not explicitly mentioned in the semi-formal presentation in [14], but tacitly used in the developments. The system of equations derived there corresponds exactly to the one derived in this presentation, only the conditions are missing.

It is interesting that the presence of these conditions occurred during the attempt to type-check a naive formalization of the development without conditions on the computer, which uncovered typing errors. This fact points to a significant advantage of the use of Deva to investigate existing methods and developments: The strict logical control (via typing) may uncover hidden assumptions and omitted proof steps.

One could ask, whether the unconditional equation system could not be derived regardless of these technical difficulties since the pruning algorithm is initialized with the unproblematic interval from $\perp$ to $\top$ and, as mentioned above, the conditions are actually invariant properties. However, this is not a Deva issue but an issue of algorithm calculation.

# 7 Conclusion

In the previous chapters we introduced the generic development language Deva and reported on two experiments in which certain aspects of formal software development methods were completely expressed in the Deva language. We now go on to discuss some of the benefits and shortcomings of these experiments and, in the light of these, to reexamine the major design choices of Deva.

## Provable Adequacy

An important question which can be asked of our Deva formalizations is whether they do indeed formalize what they claim to formalize, namely VDM and calculational developments. We wish to examine this question of adequacy from two angles: the theoretical and the pragmatic. The most satisfying answer would, of course, be a *proof* of adequacy. In fact, in Chap. 3 we showed a framework in which such proofs of adequacy could be performed, but, we have not given such proofs for our formalizations. There are two reasons why we have not done so. First, we shunned the amount of work associated with such proofs. We knew that in principle it could be done and that it has been done, at least for small select pieces of formalizations, such as predicate logic or Peano-arithmetic (cf. [30]). Second, beyond the logical basis and the data types of the development methods, it was not at all clear against what adequacy proofs should be performed, simply because the development methods were not formally defined. This situation has recently changed somewhat as a drafted VDM-standard [2] and publications like [58] have become available.

Since we have not given proof of the adequacy of our formalizations, an understanding of the Deva language and its mathematical properties must suffice for the reader to trust the formalizations. This situation is admittedly far from ideal and must be improved upon. It is, however, common to other work in this area. A reassuring fact, though, is that the properties of Deva ensure that Deva itself does not introduce new errors, i.e. any error, such as logical inconsistency, in a formalized theory must lie in the formalization itself.

## Pragmatic Adequacy

It is more important for us to assess the adequacy of our experiments from a pragmatic point of view: Did our experiments cover the pragmatic aspects of VDM and calculational development? The honest and straightforward answer is that, while essential methodological aspects were adequately expressed in Deva, the complete formalization approach made VDM and calculational development too complex and time-consuming for practical use. We illustrate this by discussing some underlying technical issues in more detail.

Deva enforces the formalization of all the necessary basic theories and the theories underlying the design methods of the case studies. This problem was tackled basically by constructing successive formal approximations in Deva and validating them with reference to the case studies. In retrospect, while the development of the basic theories involved a quite straightforward transcription of material available in good handbooks, a fairly large amount of effort went into getting the method formalizations "right". We found it much harder to develop a precise understanding of a design method as a set of theories than to use Deva to formalize these theories. In fact, it appears very important to separate two activities, at least conceptually: first, to elaborate a clear and complete theory of a method, with elegant proofs, and to express it semi-formally in a model handbook; second, to "code" this theory in Deva or in some other similar language or system. The problem was that complete theories of design methods are rarely available, so to use Deva we were forced us to develop them on our own. The effort involved, however, can be considered worthwhile because it promotes understanding of methodological issues and it results in a quite usable fragment of a general library. We believe that the scaling-up of formal developments will go hand in hand with the development of extensive and well-structured libraries of theories containing notations, laws, development schemes, and tactics and ranging from basic mathematics, to development methodologies to concrete application areas. Given the availability of such libraries, it is tempting to experiment with novel combinations of the calculi and data types underlying the different methodologies. For example, on the basis of the theories presented in this book, it is, technically, a simple matter to experiment with a variant of VDM using the calculationally oriented presentation of sequences from the second case study.

Deva enforces the explicit type declaration of each object involved in a formalization. This requirement is certainly debatable since many simple object declarations can be inferred by inspection of the kinds of uses of the objects. One can envisage an ML-like mechanism to automatically infer missing declarations. On the other hand, we quite frequently found (especially in the more complex cases) that this added type information enhanced clarity and understanding of a development. Still, the user should have the choice, and so this is another topic for further research.

Deva requires the explicit statement of all side conditions in rules and developments. From a logical point of view, this is definitely an advantage. Pragmatically, however, developments become more complex since side conditions must be discharged by auxiliary proofs before the rule itself can be applied. In our experience, the crucial challenge in formalizing developments is to identify a set of abstract, yet illuminating, laws from which all auxiliary proofs can be economically composed.

Deva requires the explicit evaluation of all syntactic operations as used by specific methods. Examples of such operations rarely occur in the formalizations presented in this book, but one example is the evaluation of selection in composite VDM objects. A more complex example would be the normalization of schemas in Z. In some such cases, one can deal with syntactic operations by defining a suitable tactic, for example, the selection tactic to evaluate selections from composite VDM objects. However, the non-recursive tactics in Deva are often too weak for that purpose. Z schema normalization cannot, for example, be evaluated by Deva tactics in a natural way. The lack of recursive tactics in Deva is probably the language's most serious expressive shortcoming. Such expressive shortcomings of languages are frequently overcome by support tools in the form of preprocessors. For instance, one can imagine a preprocessor for Devil which performs Z schema normalization. However, these tools have to be carefully designed so that error reports, for example, are given in terms of the original input, and not in terms of the translation.

Deva requires that the logical chain in developments be made fully explicit. In rigorous developments on paper this chain is usually ensured by appropriate comments in between the individual development steps (cf. the informal and the formalized version of the proof of the binomial formula). Well-presented Deva developments are, in fact, not too far from this presentation style: A comment is replaced by an application of (or a cut with) a law or a tactic.

In general, we can identify three levels on which Deva assists in the (static) presentation of formal developments and formalizations: Deva's own syntax, the possibility of defining mixfix operators, and the facilities introduced by the WEB system. The design of Deva's own syntax was mainly driven by experiments with expressing program deductions. More experiments might well suggest adaptations (see below). The possibility of declaring or defining infix, prefix, and postfix operators was essential to cover a sufficient amount of the methods' notation. There is, however, room for improvement here. For example, precedence parsing is inadequate for handling arbitrary context-free notations, such as the VDM or Squiggol notations for enumerated lists. Notations like these are, however, not in principle beyond the scope of a generic development language. After all, many generic parsing tools are available and can be used in a preprocessing step. The WEB system frees the user from the straitjacket of a formal language. We have not made full use of WEB's features, mostly because the system was not available when we developed the formalizations. Also, it is still a prototype and can be improved in several ways. One such improvement, for example, would be to control the spacing around operators depending on their height in the syntax tree. What we have not yet paid any attention to is the dynamic design and presentation of a development and its formalizations. At the moment, we do not have a sophisticated graphical editor for entering and editing Deva documents but continue to use conventional text editors. This is, of course, not what the current state-of-the-art would lead us to expect. Graphical editors such as MathPad (cf. [9]) or $G^2F$ (cf. [40] provide nice examples of what such editors might look like. Another exciting idea suggested by the web-like structure of the presentation of developments is to extend the editor by incorporating hyper-text facilities so that the users can navigate through their formalizations, zoom into a development to view its internal details or merely examine it on a more abstract level. This idea is also put forward by Lamport in his short paper on how to write proofs [69]. Such a tool should, of course, be integrated into the complete support environment. In particular, it should have a natural interface to the checker.

Within the formalizations, the degree of automation was rather low. Usually proof step details or details of reasoning chains were left implicit by synthesis of arguments through pattern-matching or by using alternatives and iterations. But there were no sophisticated induction tactics and no reasoning modulo a ruleset, such as associativity and commutativity. With regard to these very important kinds of automation, the current Deva version cannot compete with many related approaches, especially those that are tailored to a particular method. In this sense, our Deva formalizations of VDM and calculational development remain at the prototyping level. This situation could probably be improved in the future by interfacing Deva with more specialized and more convincingly automated proof support systems. We believe that the use of such systems is quite indispensable when trying to tackle more significant applications than those presented in this book. Of course, the combination of such systems with Deva poses a number of engineering, as well as theoretical, challenges.

All these technical points raise a more general question: What is the relation between performing developments rigorously and formalizing them in Deva? How much extra effort is involved when using Deva? Clearly, there is the extra effort of learning the Deva language. We hope that the present book makes this a feasible task, although we do not claim it is an easy one. Apart from that, it is very difficult to give a plausible answer to this question, because so little experimental material is yet available as evidence. Nevertheless, for the reader's interest, amusement, or disgust, we propose a rather quantitative answer: the formalization of a rigorous development in Deva leads to an expansion resulting from two sources: the internal "formal noise" of Deva, and the amount of "real" additional information needed to formalize the rigorous development.

In our experience, the first source is characterized by an expansion factor usually located within the interval $[1, 2]$; in other words, the formal noise of (the current version of) Deva rarely more than doubles the original size of a carefully presented rigorous development. We merely wish to measure the expansion factor of a Deva development relative to a rigorous development. Hence, this figure does not take into account the size of the formalizations of the necessary basic theories, which can, admittedly, be considerable. Future language developments should try to reduce the expansion factor; it should, ideally, be located within the interval $[0, 1]$ (excluding, of course, 0), i.e. formalization of developments should, ideally, lead to a reduction in size, as is the case with the formalization of algorithmic descriptions in good programming languages. Of course, these comparisons are based on the size of "mature" — i.e. repeatedly reviewed, restructured, and simplified — formalizations, and not on initial versions for martyrs.

The second source of expansion differs according to specific methodologies and their respective requirements as to how much information should be given in a rigorous development. For example, we found calculational developments to require less additional information than VDM developments. In the case of this second expansion factor, we therefore feel unable to set an upper bound valid for any method and any developer using it. Ideally, a rigorous development should already contain all information necessary for its correctness, in other words, the factor should be 1. However such program derivations are rarely available, the use of Deva forced us to develop them on our own; a problem that was already mentioned above in the context of the formalization of programming methods.

This statement about the extra effort involved when using Deva should be taken with a pinch of salt. First, the distinction between the two sources of expansion is not quite sharp, for example, because the need for additional information can often arguably be seen as a weakness of the current Deva-tactics. Second, purely quantitative measurement is clearly insufficient, it does not, for example, cover good structuring and readability.

A particularly intriguing set of future experiments might be to rewrite existing case studies and good textbooks on rigorous software design in Deva and then compare the expansion factors. Of course, such exercises should be performed parallel to the use of approaches other than Deva, e.g. those discussed in the introduction. This would allow a comparison of the expansion factors, with respect to both size and obscurity, yielded by Deva with those of related approaches.

## Review of the General Design of Deva

In our introduction, we listed a number of ideal requirements for generic development languages. We then went on to illustrate how developments can be formalized on the basis of specifific design choices made with respect to the Deva language. Previous discussions have, essentially, reviewed these results from an internal viewpoint, i.e., from inside the Deva framework. Going on from this, we now want to step outside this framework and review the general design of Deva itself. This is particularly useful since the Deva language is in a process of active evolution, and future versions of the language should, hopefully, reduce the drawbacks of the current one. In one sentence, the general design of (the current version of) Deva can be summarized as the adoption of a generalized typed $\lambda$-calculus as a basic framework and its extension to cover products, sums, compositions, theories, and implicit developments.

Adopting a typed $\lambda$-calculus allowed us to encode natural deduction-style proofs (cf. the figures on page 40 and 39) and calculational style proofs (cf. the figures on page 32 and 32). Allowing types to be $\lambda$-structured themselves effected a very economic generalization of this approach, and this provided the essential expressive power to formalize logics.

Nevertheless, there are some drawbacks to this approach: In typed $\lambda$-calculi, assumptions are recorded as *type declarations*, a technique which, essentially, amounts to introducing pointers, i.e., identifiers to formulas. If enforced, as in Deva, this pointer technique can lead to some unnecessary technical complications. For example, it would sometimes be more elegant to record assumptions directly as *sets of assumed formulas*, as is done in sequent calculus.

The bottom-up grafting of various structuring concepts on the basic $\lambda$-calculus framework was essentially driven by experiments with expressing program deductions. The cut as a key concept of compositional development, remained extremely technical in its bottom-up definition, though, ideally, composition should be as simple and basic as application.

A major drawback of our approach to implicit developments, i.e., considering implicit developments to be valid if they can be explained by explicitly valid developments, is its lack of real compositionality, i.e., the implicit validity of a constructed development is not described in terms of the implicit validity of its parts. Rather, it is described in terms of the *explanations*, i.e., compilations into explicit developments, of its parts. Moreover, our approach has led to a two-level, i.e., implicit/explicit, view of developments. Instead of such a black or white situation, in many cases one would prefer a continuum stretching from highly implicit to fully explicit developments.

One would, ideally, like to understand a development language from looking at its algebraic properties. Unfortunately, our $\lambda$-calculus-based approach led to an operational, i.e., reduction-based, language definition whose consistency rests on rather deep confluence and strong normalization theorems. It would be nice to have, at least in addition, a more algebraic specification of the essential concepts underlying Deva. Current work is elaborating an algebraic framework in which the basic ideas underlying Deva can be captured by a set of crisp algebraic laws and which avoids many of the drawbacks discussed above (see [94] for a first approximation).

## Future Work

Let us briefly summarize what we anticipate to be potential areas for future research. In order to better understand and isolate the key concepts for expressing and supporting formal software developments, more case studies must be conducted. Such case studies should guide and inspire the further development of the Deva language and its implementation. Then there are syntactic issues: What facilities are needed to express formalizations even closer to the syntax of development methods? Is the idea of using a preprocessor to translate, for example, VDM developments into Deva expressions reasonable? And there are important semantic issues: How can Deva be extended by a more powerful concept of tactics, including typed recursion? How can interfaces to (external, specific) provers be defined and related to Deva? Is there a set of algebraic laws that characterizes the key concepts of developments as identified in Deva? Finally, there are the issues of tool support: extending and adapting a library of theories for software developments, supporting the interactive development and "debugging" formalizations, implementing interfaces to external systems such as provers and preprocessors, and designing a graphical user interface with hypertext features extending the current WEB system. Thus, not too surprisingly, future research areas include logic and algebra as well as database management and interface design.

## Final Remarks

The main objective of this book was to show how the generic development language Deva can be used to express formal software developments. The long term goal is to tackle case studies of a more significant size. This can only be achieved through exchange and cooperation with related research fields such as mathematics of programming, proof languages, support systems, and specific application domains. We hope that the book will prove a helpful tool in this interaction, and we would very much like to see other approaches to formal software development published in a similar — and, of course, eventually much better — form.

# A  Machine-level Definition of the Deva Kernel

As promised in Sect. 3.2.6, we will present a formal definition of the kernel of
the explicit part of Deva using de Bruijn indices. We will proceed exactly as in
Sect. 3.2 and will point out the major differences.

## A.1  Formation

Let $\mathcal{V}_t$ denote the (denumerable) set of *text identifiers*, $\mathcal{V}_c$ the (denumerable)
set of *context identifiers*, and N the positive natural numbers, then the syntax
of the kernel of the explicit part of Deva is specified by the following grammar:

$$\mathcal{T} := \mathbf{prim} \mid \mathbf{N} \mid [\mathcal{C} \vdash \mathcal{T}] \mid \mathcal{T}(\mathbf{N} := \mathcal{T}) \mid \mathcal{T} \therefore \mathcal{T},$$
$$\mathcal{C} := \mathcal{V}_t : \mathcal{T} \mid \mathcal{V}_t := \mathcal{T} \mid \mathbf{context}\,\mathcal{V}_c := \mathcal{C} \mid \mathbf{import}\,\mathbf{N} \mid [\mathcal{C};\mathcal{C}].$$

Note that the only difference to Deva is that the *use* of a text- or a context-
identifier is now denoted by a natural number. In the following, indices are
typeset in boldface in order to make them more easily recognizable.

## A.2  Environments

Since environments are the major tool for identifying bindings, it should be
obvious that most of the changes to definition of Deva given in the main text
will involve the handling of environments. Again, an environment is a linear list
of bindings, where a binding is either a declaration, a definition, or a context
definition.

The partial evaluation of context operations being carried out while pushing a
context onto an environment is somewhat complicated to define. Since de Bruijn
indices represent the static distance to a binding occurrence, they have to be
adjusted during a partial evaluation of context operations. This adjustment
of indices is accomplished by the shift function which takes as arguments an
environment $E$, a text or a context $e$, and two integers $i$ and $k$, and returns as
a result a text or a context which is derived from $e$ by adjusting by $i$ all indices
(relative to $E$) except those smaller than or equal to $k$. As an abbreviation, we
will write shift$(e, i)$ for shift$(e, i, 0)$. The definition of the shift function is now
presented together with the definition of pushing a context onto an environment
in a mutually recursive way. First of all, in the following cases the definition of
the shift function is pretty straightforward:

$$\text{shift}_E(\mathbf{prim}, i, k) = \mathbf{prim}$$
$$\text{shift}_E(t_1(n := t_2), i, k) = \text{shift}_E(t_1, i, k)(n := \text{shift}_E(t_2, i, k))$$
$$\text{shift}_E(t_1 \therefore t_2, i, k) = \text{shift}_E(t_1, i, k) \therefore \text{shift}_E(t_2, i, k)$$

$$\text{shift}_E(x : t, i, k) = x : \text{shift}_E(t, i, k)$$
$$\text{shift}_E(x := t, i, k) = x := \text{shift}_E(t, i, k)$$
$$\text{shift}_E(\textbf{context } p := c, i, k) = \textbf{context } p := \text{shift}_E(c, i, k)$$

An index $n$ is shifted if it points behind the first $k$ bindings:

$$\text{shift}_E(n, i, k) = \begin{cases} n + i & \text{if } n > k \\ n & \text{otherwise} \end{cases}$$

$$\text{shift}_E(\textbf{import } n, i, k) = \begin{cases} \textbf{import } n + i & \text{if } n > k \\ \textbf{import } n & \text{otherwise} \end{cases}$$

In the following two situations, those indices which point to bindings contained in the context $c$ must not be shifted — this explains the need for the third argument of the shift function.

$$\text{shift}_E([c \vdash t], i, k) = [c' \vdash \text{shift}_{E \oplus c'}(t, i, k + k')]$$
$$\text{shift}_E([\![ c; c_1 ]\!], i, k) = [\![ c'; \text{shift}_{E \oplus c'}(c_1, i, k + k') ]\!]$$

where

$$c' = \text{shift}_E(c, i, k) \text{ and } k = \#(E \oplus c') - \#E.$$

Finally, we define the operation of pushing a context onto an environment. Note that the lookup function defined in the main text may now be replaced by the expression $E.n$ denoting the $n$-th entry (from the right) in the environment $E$. In case this entry does not exists (because $n$ is too large), $E.n$ is undefined. This should not cause major concern since, in general, all the definitions given in this text make sense only if they are closed relative to some environment.

$$E \oplus x : t = E \oplus \langle x : t \rangle,$$
$$E \oplus x := t = E \oplus \langle x := t \rangle,$$
$$E \oplus \textbf{context } p := c = E \oplus \langle \textbf{context } p := c \rangle,$$
$$E \oplus \textbf{import } n = E \oplus \text{shift}_E(c, n), \quad \text{if } E.n = \textbf{context } p := c,$$
$$E \oplus [\![ c_1; c_2 ]\!] = (E \oplus c_1) \oplus c_2.$$

## A.3  The definition of $\tau$ and $\kappa$

The definition of the bijection (restricted to closed texts and contexts) $\tau$ which translates relative to an environment a text or a context using concrete names to the corresponding text or context using de Bruijn's indexing scheme is given below:

$$\tau_E(\mathbf{prim}) = \mathbf{prim}$$
$$\tau_E(x) = \mathrm{pos}_E(x)$$
$$\tau_E([c \vdash t]) = [\tau_E(c) \vdash \tau_{E \oplus c}(t)]$$
$$\tau_E(t_1(n := t_2)) = \tau_E(t_1)(n := \tau_E(t_2))$$
$$\tau_E(t_1 \therefore t_2) = \tau_E(t_1) \therefore \tau_E(t_2)$$
$$\tau_E(x : t) = x : \tau_E(t)$$
$$\tau_E(x := t) = x := \tau_E(t)$$
$$\tau_E(\mathbf{context}\ p := c) = \mathbf{context}\ p := \tau_E(c)$$
$$\tau_E(\mathbf{import}\ p) = \mathbf{import}\ \mathrm{pos}_E(p)$$
$$\tau_E(\llbracket c_1; c_2 \rrbracket) = \llbracket \tau_E(c_1); \tau_{E \oplus c_1}(c_2) \rrbracket$$

Here, $\mathrm{pos}_E(x)$ denotes the *position* of the first link describing $x$ counting from the right. Again, this operation may be undefined if $x$ is not closed relative to $E$.

The definition of $\kappa$ is trivial since $\kappa$ just steps recursively through its argument and replaces any identifier at a binding by the standard identifier '$\sqcup$'. We only give the following three equations since in all other situations $\kappa$ is applied recursively to the components of its argument.

$$\kappa(x : t) = \sqcup : \kappa(t)$$
$$\kappa(x := t) = \sqcup := \kappa(t)$$
$$\kappa(\mathbf{context}\ p := c) = \mathbf{context}\ \sqcup := \kappa(c)$$

## A.4  Closed texts and closed contexts

The definition of closed texts and closed contexts is almost exactly the same as in Sect. 3.2.4. Of course identifiers have to be replaced by indices and the crucial two rules which check whether an identifier is defined within the environment are replaced by

$$\frac{\vdash_{cl} E \qquad E.n \text{ is defined}}{E \vdash_{cl} n} \qquad \text{and} \qquad \frac{\vdash_{cl} E \qquad E.n \text{ is defined}}{E \vdash_{cl} \mathbf{import}\ n}$$

## A.5  Reduction of texts and contexts

Reduction is defined similarly to the definition given in Sect. 3.2.5. But now, one has to be careful about possible adjustment of indices during the reduction process. Altogether, except for the obvious changes, the following axioms replace their respective counterparts:

$$E \vdash n \ \triangleright \ \text{shift}_E(t, n), \quad \text{if } E.n = x := t,$$
$$E \vdash [x : t_1 \vdash t_2](1 := t) \ \triangleright \ [x := t \vdash t_2],$$
$$E \vdash [x : t_1 \vdash t_2](n + 1 := t) \ \triangleright \ [x : t_1 \vdash t_2(n := \text{shift}_E(t, 1))],$$
$$E \vdash [x := t_1 \vdash \text{shift}_E(t_2, 1)] \ \triangleright \ t_2$$
$$E \vdash [\textbf{context } p := c \vdash \text{shift}_E(t, 1)] \ \triangleright \ t$$

$$E \vdash \textbf{import } n \ \triangleright \ \text{shift}_E(c, n), \quad \text{if } E.n = \textbf{context } p := c.$$

## A.6    Conversion

The rules for conversion are obvious and are thus omitted.

## A.7    Type assignment

The rules for type assignment are exactly the same as in Sect. 3.2.7 except for
the following:

$$\text{typ}_E(n) = \begin{cases} \text{shift}_E(t, n) & \text{if } E.n = x : t \\ \text{typ}_E(\text{shift}(t, n)) & \text{if } E.n = x := t \\ \text{undefined} & \text{if } E.n = \text{undefined} \end{cases}$$

## A.8    Auxiliary functions and predicates for validity

The name-irrelevant conversion is defined as in Sect. 3.2.8. As for the definition
of the 'domain' relation, only the following rules need to be adjusted:

$$\frac{E \vdash \text{typ}_E(t_1) \ \sim \ [x : t_2 \vdash t_3] \qquad n\text{-dom}_{E \oplus x : t_2}(\text{shift}(t_1, 1)(1 := x), \text{shift}_E(t_4, 1)) \quad E \vdash_{cl} t_4}{(n + 1)\text{-dom}_E(t_1, t_4)}$$

$$\frac{E \vdash t_1 \ \sim \ [x : t_2 \vdash t_3] \qquad n\text{-dom}_{E \oplus x : t_2}(t_3, \text{shift}_E(t_4, 1)) \quad E \vdash_{cl} t_4}{(n + 1)\text{-dom}_E(t_1, t_4)}$$

## A.9    Validity

The same changes apply to the definition of validity as for the definition of closed
texts or contexts.

# B    Index of Deva Constructs

# C   Crossreferences

## C.1   Table of Deva Sections Defined in the Tutorial

## C.2   Index of Variables Defined in the Tutorial

## C.3    Table of Deva Sections Defined in the Case Studies

⟨ $\alpha\beta$-pruning. 6.7.13, 6.7.14, 6.7.15, 6.7.18, 6.7.19, 6.7.20, 6.7.21, 6.7.22, 6.7.24 ⟩    This code is used in section 6.7.

⟨ Abstract initialization. 5.5.6.13 ⟩    This code is used in section 5.5.6.12.

⟨ Abstract version verification. 5.5.6.10 ⟩    This code is used in section 5.5.6.5.

⟨ Arbitrary branching trees. 6.7.1 ⟩    This code is used in section 6.7.

⟨ Assembly of main verification. 5.5.6.5 ⟩    This code is used in section 5.5.6.

⟨ Auxiliary Deductions. 5.5.7.3 ⟩    This code is used in section 5.5.7.2.

⟨ Axioms of finite maps. 4.3.5.2 ⟩    This code is used in section 4.3.5.

⟨ Axioms of finite sets. 4.3.2.2 ⟩    This code is used in section 4.3.2.

⟨ Axioms of join lists. 6.2.2 ⟩    This code is used in section 6.2.

⟨ Axioms of natural numbers. 4.3.1.2 ⟩    This code is used in section 4.3.1.

⟨ Axioms of non-empty join lists. 6.3.2 ⟩    This code is used in section 6.3.

⟨ Axioms of propositional logic. 4.2.1.3 ⟩    This code is used in section 4.2.1.1.

⟨ Axioms of sequences. 4.3.3.2 ⟩   This code is used in section 4.3.3.

⟨ Axioms of tuples. 4.3.4.2 ⟩   This code is used in section 4.3.4.

⟨ Base case. 5.6.6.3 ⟩   This code is used in section 5.6.6.2.

⟨ Basic theories for algorithm calculation. 4.1.3 ⟩   This code is used in section 4.1.

⟨ Basic theories of VDM. 4.1.2 ⟩   This code is used in section 4.1.

⟨ Body of the proof of the second retrieve lemma. 5.5.5.1 ⟩   This code is used in section 5.5.5.

⟨ Boolean algebra. 4.4.3.5, 4.4.3.6, 4.4.3.7, 4.4.3.8 ⟩ This code is used in section 4.4.3.1.

⟨ Calculation of $\text{leval}_\beta^\beta \circ \text{fork}$. 6.7.17 ⟩   This code is used in section 6.7.15.

⟨ Calculation of $\text{leval}_\alpha^\beta \circ \text{tip}$. 6.7.16 ⟩   This code is used in section 6.7.15.

⟨ Case of negative reaction. 5.5.8.7 ⟩   This code is used in section 5.5.8.4.

⟨ Case of positive reaction (not checked). 5.5.8.5 ⟩

⟨ Case of positive reaction. 5.5.8.6 ⟩   This code is used in section 5.5.8.4.

⟨ Catamorphisms (for non-empty join lists). 6.5.2 ⟩   This code is used in section 6.5.

⟨ Catamorphisms. 6.4.2 ⟩   This code is used in section 6.4.

⟨ Concrete initialization. 5.5.6.9 ⟩   This code is used in section 5.5.6.8.

⟨ Concrete version verification. 5.5.6.6 ⟩   This code is used in section 5.5.6.5.

⟨ Construction of the abstract version. 5.5.2.8 ⟩   This code is used in section 5.5.2.9.

⟨ Construction of the concrete version. 5.5.3.3 ⟩   This code is used in section 5.5.3.4.

⟨ Construction of versions. 5.3.2.1 ⟩   This code is used in section 5.3.2.4.

⟨ Constructors of finite maps. 4.3.5.1 ⟩   This code is used in section 4.3.5.

⟨ Constructors of finite sets. 4.3.2.1 ⟩   This code is used in section 4.3.2.

⟨ Constructors of join lists. 6.2.1 ⟩   This code is used in section 6.2.

⟨ Constructors of natural numbers. 4.3.1.1 ⟩   This code is used in section 4.3.1.

⟨ Constructors of non-empty join lists. 6.3.1 ⟩   This code is used in section 6.3.

⟨ Constructors of sequences. 4.3.3.1 ⟩   This code is used in section 4.3.3.

⟨ Constructors of tuples. 4.3.4.1 ⟩   This code is used in section 4.3.4.

⟨ Declaration of the other abstract operations. 5.5.2.7 ⟩   This code is used in section 5.5.2.9.

⟨ Declaration of the other concrete operations. 5.5.3.2 ⟩   This code is used in section 5.5.3.4.

⟨ Definition of VDM reification. 5.3.3.5 ⟩   This code is used in section 5.3.3.6.

⟨ Definition of the retrieve function. 5.5.4.1 ⟩   This code is used in section 5.5.4.6.

⟨ Derivation of selection laws. 5.5.2.1 ⟩   This code is used in section 5.5.2.9.

⟨ Derived laws of propositional logic. 4.2.1.6 ⟩   This code is used in section 4.2.1.1.

⟨ Derived operators and laws of finite sets. 4.3.2.3, 4.3.2.4, 4.3.2.5, 4.3.2.6, 4.3.2.7, 4.3.2.8, 4.3.2.9 ⟩   This code is used in section 4.3.2.

⟨ Derived operators and laws of sequences. 4.3.3.3, 4.3.3.4, 4.3.3.5, 4.3.3.6, 4.3.3.7 ⟩ This code is used in section 4.3.3.

⟨ Derived operators of finite maps. 4.3.5.3, 4.3.5.4 ⟩   This code is used in section 4.3.5.

⟨ Derived operators of natural numbers. 4.3.1.3 ⟩   This code is used in section 4.3.1.

⟨ Derived operators of propositional logic. 4.2.1.4 ⟩   This code is used in section 4.2.1.1.

⟨ Derived properties of tuples. 4.3.4.4, 4.3.4.5, 4.3.4.6 ⟩   This code is used in section 4.3.4.

⟨ Distributive monoid pair. 4.4.3.4 ⟩   This code is used in section 4.4.3.1.

⟨ Empty sort. 5.3.1.2 ⟩   This code is used in section 5.3.1.4.

⟨ Extensional equality of terms or functions. 4.4.1.1 ⟩   This code is used in section 4.1.3.

⟨ Finite maps. 4.3.5 ⟩   This code is used in section 4.1.2.

⟨ Finite sets. 4.3.2 ⟩   This code is used in section 4.1.2.

⟨ First block. 5.6.4.3 ⟩   This code is used in section 5.6.4.2.

⟨ First explicitation of the development of mss (not checked). 6.6.6 ⟩

⟨ First five abstract operations. 5.5.6.11 ⟩   This code is used in section 5.5.6.10.

⟨ First five concrete operations. 5.5.6.7 ⟩   This code is used in section 5.5.6.6.

⟨ First five operations. 5.5.6.15 ⟩   This code is used in section 5.5.6.14.

⟨ First three abstract operations. 5.5.6.12 ⟩   This code is used in section 5.5.6.11.

⟨ First three concrete operations. 5.5.6.8 ⟩   This code is used in section 5.5.6.7.

⟨ First three operations. 5.5.6.16 ⟩   This code is used in section 5.5.6.15.

⟨ Fold rules for the overloaded equality. 4.4.1.5 ⟩   This code is used in section 4.4.1.1.

⟨ Frequently used contexts. 5.6.1 ⟩   This code is used in section 5.6.3.2.

⟨ Further axioms and laws of predicate logic. 4.2.1.8, 4.2.1.9, 4.2.1.10, 4.2.1.11, 4.2.1.12 ⟩   This code is used in section 4.2.1.7.

⟨ General evaluation tactic for the HLA case study. 5.5.4.4 ⟩   This code is used in section 5.5.4.6.

⟨ Generic algebraic properties. 4.4.3.2 ⟩   This code is used in section 4.4.3.1.

⟨ HLA abstract specification of the state and the invariant. 5.5.2 ⟩   This code is used in section 5.5.2.9.

⟨ HLA abstract specification. 5.5.2.9 ⟩   This code is used in section 5.5.

⟨ HLA concrete specification of the state and the invariant. 5.5.3 ⟩   This code is used in section 5.5.3.4.

⟨ HLA concrete specification. 5.5.3.4 ⟩   This code is used in section 5.5.

⟨ HLA global parameters. 5.5.1 ⟩   This code is used in section 5.5.

⟨ HLA retrieve function. 5.5.4.6 ⟩   This code is used in section 5.5.

⟨ HLA verification. 5.5.6 ⟩   This code is used in section 5.5.

⟨ Horner's rule. 6.6.2 ⟩   This code is used in section 6.6.

⟨ Human leukocyte antigen typing problem. 5.5 ⟩   This code is used in section 5.1.

⟨ Implicit development of mss (not checked). 6.6.5 ⟩   This code is used in section 6.6.4.

⟨ Imports needed by *CalculationalBasics*. 4.4.1 ⟩   This code is used in section 4.1.3.

⟨ Imports needed by *VDMBasics*. 4.3 ⟩   This code is used in section 4.1.2.

⟨ Induced Partial Ordering. 4.4.4 ⟩   This code is used in section 4.1.3.

⟨ Induction on version triples. 5.6.2.4 ⟩   This code is used in section 5.6.3.2.

⟨ Induction on versions. 5.3.2.3 ⟩   This code is used in section 5.3.2.4.

⟨ Induction principle. 5.6.2.1 ⟩   This code is used in section 5.6.2.4.

⟨ Inductive proof of the transitivity condition. 5.6.6.2 ⟩   This code is used in section 5.6.6.1.

⟨ Inductive step. 5.6.6.4 ⟩   This code is used in section 5.6.6.2.

⟨Specification of *DetPosNeg*. 5.5.2.6⟩   This code is used in section 5.5.2.9.
⟨Specification of *cDetPosNeg*. 5.5.3.1⟩   This code is used in section 5.5.3.4.
⟨Tactic for evaluating the retrieve function. 5.5.4.3⟩   This code is used in section 5.5.4.6.
⟨Terms involving functions. 4.4.2.1⟩   This code is used in section 4.1.3.
⟨The "maximum segment sum" problem. 6.6.4⟩   This code is used in section 6.6.
⟨Third explicitation of the development of mss. 6.6.8⟩
⟨Transitivity of operator reification. 5.6.5.1⟩   This code is used in section 5.6.3.4.
⟨Transitivity of reification. 5.6.3.2⟩   This code is used in section 5.1.
⟨Transitivity of the reification condition. 5.6.6.1⟩   This code is used in section 5.6.3.4.
⟨Tree evaluation problems. 6.7⟩   This code is used in section 6.1.
⟨Tuples. 4.3.4⟩   This code is used in section 4.1.2.
⟨Two auxiliary specifications. 5.5.4⟩   This code is used in section 5.5.4.6.
⟨Two retrieve lemmas. 5.5.4.5⟩   This code is used in section 5.5.4.6.
⟨Unfold rules for the overloaded equality. 4.4.1.4⟩   This code is used in section 4.4.1.1.
⟨VDM operations. 5.3.1.4⟩   This code is used in section 5.3.
⟨VDM reification. 5.3⟩   This code is used in section 5.1.
⟨VDM tactics. 4.3.6⟩   This code is used in section 4.1.2.
⟨VDM version reification. 5.3.3.6⟩   This code is used in section 5.3.
⟨VDM versions. 5.3.2.4⟩   This code is used in section 5.3.
⟨Verification of the retrieve condition. 5.6.4.1⟩   This code is used in section 5.6.3.4.
⟨Versions and operations. 5.6.1.5⟩   This code is used in section 5.6.1.
⟨Versions and retrieve functions. 5.6.1.6⟩   This code is used in section 5.6.1.
⟨Versions. 5.6.1.3⟩   This code is used in section 5.6.1.
⟨*Fail* selection. 5.5.2.5⟩   This code is used in section 5.5.2.1.
⟨*Neg* selection. 5.5.2.4⟩   This code is used in section 5.5.2.1.
⟨*Pos* selection. 5.5.2.3⟩   This code is used in section 5.5.2.1.
⟨*Resexp* selection. 5.5.2.2⟩   This code is used in section 5.5.2.1.

## C.4   Index of Variables Defined in the Case Studies

# D References

1. J. R. Abrial. The B-Tool (abstract). In R. Bloomfield, L. Marshall, and C. Jones, editors, *VDM'88 - The Way Ahead*. Springer-Verlag, 1988.
2. D. Andrews. VDM specification language: Proto-Standard. Draft proposal, BSI IST/5/50, December 1992.
3. M. Anlauff. *A Support Environment for a Generic Devlopment Language*. Forthcoming dissertation., TU Berlin, 1994.
4. R. D. Arthan. On formal specification of a proof tool. In *VDM'91 Formal Software Developments*, volume 551 of *LNCS* , pages 356–370, 1991.
5. A. Avron, F. Honsell, I. A. Mason, and R. Pollack. Using typed $\lambda$-calculus to implement formal systems on a machine. *Journal of Automated Reasoning*, 9(3):309–354, 1992.
6. R. C. Backhouse. Making formality work for us. *EATCS Bulletin*, 38:219–249, June 1989.
7. R. C. Backhouse, P. J. de Bruin, P. Hoogendijk, G. Malcolm, T. S. Voermans, and J. van der Woude. Polynomial relators. In *Proceedings of the 2nd Conference on Algebraic Methodology and Software Technology, AMAST '91*, Workshops in Computing, pages 303–362. Springer-Verlag, 1992.
8. R. C. Backhouse, P. de Bruin de, G. Malcolm, T. S. Voermans, and J. van der Woude. Relational catamorphisms. In Möller B., editor, *Proceedings of the IFIP TC2/WG2.1 Working Conference on Constructing Programs*, pages 287–318. Elsevier Science Publishers B.V., 1992.
9. R. C. Backhouse, R. Verhoeven, and O. Weber. Mathpad. Technical report, Technical University of Eindhoven, Department of Computer Science, 1993.
10. H. Barendregt. Introduction to generalised type systems. *Journal of Functional Programming*, 1(4):375–416, 1991.
11. H. Barringer, J. H. Cheng, and C. B. Jones. A logic covering undefinedness in programs. *Acta Informatica*, 5:251–259, 1984.
12. M. Beyer. Specification of a LEX-like scanner. Forthcoming technical report., TU Berlin, 1993.
13. M. Biersack, R. Raschke, and M. Simons. DVWEB: A web for the generic development language Deva. Forthcoming technical report., TU Berlin, 1993.
14. R. Bird. Lectures on constructive functional programming. In M. Broy, editor, *Constructive Methods in Computer Science*, volume F69 of *NATO ASI Series*, pages 151–216. Springer-Verlag, 1989.
15. R. Bird and O. de Moor. List partitions. *Formal Aspects of Computing*, 5(1):255–279, 1993.
16. *Second Workshop on Logical Frameworks*, 1992. Preliminary proceedings.
17. S. Brien and J. Nicholls. Z base standard (version 1.0). Oxford University Computing Laboratory, Programming Research Group, November 1992.
18. M. Broy and C. B. Jones, editors. *Programming Concepts and Methods*. North Holland, 1990.
19. L. Cardelli and G. Longo. A semantic basis for Quest. Research report 55, Digital, System Research Center, Palo Alto, 1990.
20. J. Cazin, P. Cros, R. Jacquart, M. Lemoine, and P. Michel. Construction and reuse of formal program developments. In *Proceedings Tapsoft 91*, LNCS 494. Springer-Verlag, 1991.
21. P. Cazin and P. Cros. Several play and replay scenarii for the HLA program development. Technical Report RR.T1-89.d, CERT, Toulouse, 1989.

22. R. Constable et al. *Implementing Mathematics with the NuPRL Proof Development System*. Prentice Hall, 1986.

23. C. Coquand. A proof of normalization for simply typed λ-calculus written in ALF. In BRA Logical Frameworks [16]. Preliminary proceedings.

24. T. Coquand and G. Huet. The calculus of constructions. *Information and Computation*, 76:95–120, 1988.

25. P. Cros et al. HLA problem oriented specification. Technical Report RR.T3-89c, CERT, Toulouse, 1989.

26. N. G. de Bruijn. Lambda calculus notation with nameless dummies. *Indigationes Mathematicae*, 34:381–392, 1972.

27. N. G. de Bruijn. A survey of the project AUTOMATH. In *To H. B. Curry: Essays on Combinatory Logic, Lambda Calculus and Formalism*, pages 579–606. Academic Press, 1980.

28. N. G. de Bruijn. Generalizing AUTOMATH by means of a lambda-typed lambda calculus. In *Prceedings of the Maryland 1984-1985 Special Year in Mathematical Logic and Theoretical Computer Science*, 1985.

29. N. G. de Bruijn. A plea for weaker frameworks. In Huet and Plotkin [54], pages 123–140.

30. P. de Groote. *Définition et Properiétés d'un Métacalcul de Représentation de Théories*. PhD thesis, University of Louvain, 1990.

31. P. de Groote. *Nederpelt's Calculus Extended with a notion of Context as a Logical Framework*, pages 69–88. Cambridge University Press, 1991.

32. P. de Groote. The conservation theorem revisited. In *Conference on Typed λ-Calculi and Applications*, volume 664 of *LNCS*, pages 163–174. Springer-Verlag, 1993.

33. P. de Groote. Defining λ-typed λ-calculi by axiomatizing the typing relation. In *STACS '93*, volume 665 of *LNCS*, pages 712–723. Springer-Verlag, 1993.

34. E. Dijkstra and W. H. J. Feijen. *A Method of Programming*. Addison-Wesley, 1988.

35. E. W. Dijkstra and C. Scholten. *Predicate Calculus and Predicate Transformers*. Springer-Verlag, 1990.

36. G. Dowek et al. The Coq proof assitant user's guide. Technical report, INRIA Rocquencourt, 1991.

37. R. Dulbecco, editor. *Encyclopedia of Human Biology*, volume 4. Academic Press, 1991.

38. A. Felty and D. Miller. Encoding a dependent-type λ-calculus in a logic programming language. In *Proceedings of CADE 1990*, volume 449 of *LNCS*, pages 221–236. Springer-Verlag, 1990.

39. B. Fields and Morten Elvang-Gøransson. A VDM case study in mural. *IEEE Transactions on Software Engineering*, 18(4):279–295, 1992.

40. R. Gabriel. The automatic generation of graphical user-interfaces. In *System design: concepts, methods and tools*, pages 589–606. IEEE Computer Society Press, 1988.

41. R. Gabriel, editor. *ESPRIT Project ToolUse, Final Report of the Deva Support Task: Retrospective and Manuals*, number 425 in "Arbeitpapiere der GMD". Gesellschaft für Mathematik und Datenverarbeitung, 1990.

42. S. J. Garland and J. V. Guttag. An overview of LP, the Larch prover. In *Proceedings of the Third International Conference on Rewriting Techniques and Applications*, volume 355 of *LNCS*, pages 137–151. Chapel Hill, N.C., 1989.

43. J.-Y. Girard, P. Taylor, and Y. Lafont. *Proof and Types*. Cambridge University Press, 1989.

44. M. J. Gordon, R. Milner, and C. P. Wadsworth. *Edinburgh LCF: A Mechanized Logic of Computation*, volume 78 of *LNCS* . Springer-Verlag, 1979.

45. M. J. C. Gordon. HOL: A proof generating system for higher-order logic. In G. Birthwhistle and P. A. Subrahmanyam, editors, *VLSI specification, Verification and Synthesis*. Kluwer, 1987.

46. M. J. C. Gordon and T. F. Melham, editors. *Introduction to HOL: A Theorem Proving Environment*. Cambridge University Press, 1993. to appear.

47. David Gries. *The Science of Programming*. Springer-Verlag, 1981.

48. J. Guttag and J. Horning. *Larch: Languages and Tools for Formal Specification*. Springer-Verlag, 1993.

49. T. Hagino. A typed lambda calculus with categorical type constructors. In D. H. Pitt, A. Poigné, and D. E. Rydeheard, editors, *Category Theory and Computer Science*, LNCS 283, pages 140–157. Springer-Verlag, 1987.

50. R. Harper, F. Honsell, and G. Plotkin. A framework for defining logics. In *Proceedings of the Symposium on Logic in Computer Science*, pages 194–204. IEEE, 1987.

51. M. Heisel, W. Reif, and W. Stephan. Formal software development in the KIV system. In M. R. Lowry and R. D. McCartney, editors, *Automating Software Design*, pages 547–576. The MIT Press, 1991.

52. J. R. Hindley and J. P. Seldin. *Introduction to Combinators and λ-Calculus*. Cambridge University Press, 1986.

53. G. Huet, editor. *Logical Foundations of Functional Programming*. Addison-Wesley, 1990.

54. G. Huet and G. Plotkin, editors. *Proceedings of the First Workshop on Logical Frameworks*. Cambridge University Press, 1991.

55. Special issue on formal methods. *IEEE Software*, September 1990.

56. B. Jacobs and T. Melham. Translating dependent type theory in higher-order logic. In *Conference on Typed λ-Calculi and Applications*, volume 664 of *LNCS* , pages 209–229. Springer-Verlag, 1993.

57. S. Jähnichen and R. Gabriel. ToolUse: A uniform approach to formal program development. *Technique et Science Informatiques*, 9(2), 1990.

58. J. Jeuring. *Theories of Algorithm Calculation*. PhD thesis, University of Utrecht, 1993.

59. C. B. Jones. *Systematic Software Development using VDM*. Prentice Hall, 1986.

60. C. B. Jones. *Systematic Software Development using VDM, second edition*. Prentice Hall, 1990.

61. C. B. Jones, K. D. Jones, P. A. Lindsay, and R. Moore. *Mural: A Formal Development Support System*. Springer, 1991.

62. S. L. Peyton Jones. *The Implementation of Functional Programming Languages*. Prentice Hall International, 1987.

63. D. Knuth. Literate programming. *The Computer Journal*, 27(2):97–111, May 1984.

64. D. Knuth. *Literate Programming*. Center for the Study of Language and Information, 1992.

65. G. Koletsos. Sequent calculus and partial logic. Master's thesis, Manchester University, 1976.

66. C. Lafontaine. Writing tactics in Deva.1: Play and replay of VDM proof obligations. Technical Report RR89-9, Université de Louvain, 1989.

67. C. Lafontaine. Formalization of the VDM reification in the Deva meta-calculus. In Broy and Jones [18], pages 333–368.

68. C. Lafontaine, Y. Ledru, and P. Schobbens. Two approaches towards the formalisation of VDM. In D. Bjorner, C. Hoare, and H. Langmaack, editors, *Proceedings of VDM'90: VDM and Z*, volume 428 of *LNCS*, pages 370–398. Springer, 1990.

69. L. Lamport. How to write a proof. Technical Report 94, DEC Systems Research Center, 1993.

70. M. Leeser. Using NuPRL for the verification and synthesis of hardware. In C. A. R. Hoare and M. C. J. Gordon, editors, *Mechanized Reasoning and Design*, pages 49–68. Prentice Hall International, 1992.

71. M. H. Liégeois. Development and replay of the HLA typing case study using VDM. Technical Report RR89-22, Université de Louvain, 1989.

72. Z. Luo. *An Extended Calculus of Constructions*. PhD thesis, University of Edinburgh, 1990.

73. Z. Luo. Program specification and data refinement in type theory. In *TAPSOFT '91*, volume 493 of *LNCS*, pages 143–168. Springer-Verlag, 1991.

74. Z. Luo and R. Pollack. The LEGO proof development system: A user's manual. Technical Report ECS-LFCS-92-211, University of Edinburgh, LFCS, 1992.

75. L. Magnusson. The new implementation of ALF. In BRA Logical Frameworks [16]. Preliminary proceedings.

76. G. Malcolm. Data structures and program transformation. *Science of Computer Programming*, 14:255–279, 1990.

77. L. Marshall. Using B to replay VDM proof obligations. Technical Report RR 87-30, Université Catholique de Louvain, 1987.

78. L. Meertens. Algorithmics: Towards programming as a mathematical activity. In J. W. de Bakker, M. Hazewinkel, and J. K. Lenstra, editors, *Proceedings of the CWI Symposium on Mathematics and Computer Science*, volume 1, pages 289–334. North Holland, 1986.

79. L. Meertens. Variations on trees. International Summer School on Constructive Algorithmics, September 1989.

80. R. P. Nederpelt. An approach to theorem proving on the basis of a typed lambda calculus. In W. Bibel and R. Kowalski, editors, *5th Conference on Automated Deduction*, volume 87 of *LNCS*, pages 182–194. Springer, 1980.

81. P. A. J. Noel. Experimenting with isabell in zf set theory. *Journal of Automated Reasoning*, 10(1):15–58, 1993.

82. L. C. Paulson. *Logic and Computation*. Cambridge University Press, 1987.

83. L. C. Paulson. The foundation of a generic theorem prover. *Journal of Automated Reasoning*, 5:363–397, 1989.

84. P. Pepper. A simple calulus for program transformation (inclusive of induction). *Science of Computer Programming*, 9(3):221–262, 1987.

85. F. Pfenning. Logic programming in the LF logical framework. In Huet and Plotkin [54], pages 149–181.

86. F Pfenning. A proof of the church-rosser theorem and its representation in a logical framework. Technical Report CMU-CS-92-186, Carnegie Mellon University, 1992.

87. R. Pollack. Implicit syntax. In *First Workshop on Logical Frameworks*, 1990. Preliminary proceedings.

88. D. Prawitz. *Natural Deduction*. Almquist & Wiskell, Stockholm, 1965.

89. T. Santen. Formalization of the SPECTRUM methodology in Deva: Signature and logical calculus. Technical Report 93-04, TU Berlin, 1993.

90. M. Simons. Basic contexts. internal communication, GMD Karlsruhe,, 1990.

91. M. Sintzoff. Suggestions for composing and specifying program design decisions. In B. Robinet, editor, *Proc. 4th Symposium on Programming*, volume 83 of *LNCS*, pages 311–326, 1980.

92. M. Sintzoff. Understanding and expressing software construction. In P. Pepper, editor, *Program Transformations and Programming Environments*, pages 169–180. Springer-Verlag, 1980.

93. M. Sintzoff. Expressing program developments in a design calculus. In M. Broy, editor, *Logic of Programming and Calculi of Discrete Design*, volume F36 of *NATO ASI Series*, pages 343–356. Springer-Verlag, 1986.

94. M. Sintzoff. Endomorphic typing. In B. Möller, H. Partsch, and S. Schuman, editors, *Formal Program Development*. Springer-Verlag, 1993. to appear.

95. M. Sintzoff, M. Weber, P. de Groote, and J. Cazin. Definition 1.1 of the generic development language Deva. ToolUse research report, Unité d'Informatique, Université Catholique de Louvain, Belgium, 1989.

96. D. R. Smith. KIDS – a knowledge based software development system. In M. R. Lowry and R. D. McCartney, editors, *Automating Software Design*, pages 483–514. The MIT Press, 1991.

97. Special issue on formal methods: part 1. *The Computer Journal*, 35(5), 1992.

98. Special issue on formal methods: part 2. *The Computer Journal*, 35(6), 1992.

99. L. S. van Bethem Jutting. *Checking Landau's "Grundlagen" in the Automath system*. PhD thesis, Eindhoven Technical University, 1979.

100. D. T. van Daalen. *The Language Theory of* AUTOMATH. PhD thesis, Technische Hoogeschool Eindhoven, 1980.

101. M. Weber. Explaining implicit proofs by explicit proofs. Private communication.

102. M. Weber. Formalization of the Bird-Meertens algorithmic calculus in the Deva meta-calculus. In Broy and Jones [18], pages 201–232.

103. M. Weber. Deriving transitivity of VDM reification in the Deva meta-calculus. In S. Prehn and W. J. Toetenel, editors, *VDM'91 Formal Software Development Methods*, volume 551 of *LNCS*, pages 406–427. Springer, 1991.

104. M. Weber. *A Meta-Calculus for Formal System Development*. Oldenbourg Verlag, 1991.

105. M. Weber. Definition and basic properties of the Deva meta-calculus. *Formal Aspects of Computing*, 1993. to appear.

# Lecture Notes in Computer Science

For information about Vols. 1–665
please contact your bookseller or Springer-Verlag